An Erie Railway train is testing the Wallkill Valley's deck truss bridge over Rondout Creek near Rosendale, New York, on the day of its official opening in 1872. Note the crowd of stovepipe-hatted officials at the upper left, watching the test. Others are standing on the locomotive and the bridge supports. Note Lock No. 7 of Delaware & Hudson Canal next to Rondout Creek. — GERALD M. BEST COLLECTION

The Beauty of

In North America - Then and Now

Richard J. Cook

Golden West Books
San Marino, California • 91108-8250

THE BEAUTY OF RAILROAD BRIDGES

... In North America — Then and Now

Published by Golden West Books
San Marino, California 91108 U.S.A.
Library of Congress Catalog Card No. 87-11998

Library of Congress Cataloging-in-Publication Data

Cook, Richard J.
The beauty of railroad bridges in North America, then and now.
Bibliography: p.
Includes index.
1. Railroad bridges—United States. 2. Railroad bridges—Canada. I. Title.
TG23.C66 1987 624'.2'0973 87-11998
ISBN 0-87095-097-5

TITLE PAGE ILLUSTRATION

A lofty concrete arch of the mighty Tunkhannock Viaduct soars majestically over the small town of Nicholson, Pennsylvania. This structure carried the main line rails of the Delaware, Lackawanna & Western across Tunkhannock Creek. — RICHARD J. COOK

Golden West Books

P.O. Box 80250 • San Marino, California • 91108-8250

It so happens that the work which is likely to be our most durable monument, and to convey some knowledge of us to the most remote posterity, is a work of bare utility; not a shrine, not a fortress, not a palace, but a bridge.

—Montgomery Schuyler
in *Harper's Weekly,* May 24, 1883

Northern Pacific No. 1527, a 2-8-2, rumbles across a beautiful wood trestle as it works an eastbound local freight at Cottonwood, Idaho, on the Camas Prairie Railroad. This road was organized in 1909 to operate as a joint line of the Union Pacific and the Northern Pacific, on a stretch of railroad between Riparia, Washington, and Grangeville, Idaho. The tracks from the former point to Lewiston (71 miles) had been built by the Oregon Railroad & Navigation Co. (a UP subsidiary), and the continuation of the line to Grangeville (79 miles) constitutes the Lapwai branch of the NP. This resulted in a 150-mile main line plus several branches and many spectacular wood trestles. The Camas Prairie is now a Union Pacific-Burlington Northern property and leases equipment from its two owners. — HENRY R. GRIFFITHS

Table of Contents

164
164

Foreword

This volume was designed for those who, like the author, find bridges a fascinating subject, especially when those bridges carry railroads. This is a book for laymen, written by a non-professional who loves trains and all things connected with the subject, such as bridges, and who does not profess to know that much about their engineering.

Rather, this is a book which, I hope, will celebrate the beauty of railroad bridges and their ability to combine that symmetry with a necessary practicability: the movement of the restless train.

Today, few railroad bridges are being erected, just as few railroads are being built. When a modern railroad bridge *is* constructed, it is usually not a major bridge, but rather a steel deck girder span or a concrete structure, often not particularly a thing of beauty.

The building of our major railway bridges coincided with the development of the railways themselves. Early major bridges were erected quickly and they were succeeded by great steel spans, truss or steel arch type or, in a few cases, by great, yet graceful, concrete structures. A few magnificent stone bridges, built so solidly in those early years, have survived and are still in use today, requiring little reinforcing. They have performed a noble service and historians have sometimes neglected to recognize their value to mankind.

This, then, is a pictorial salute to the designers and builders of these beautiful utilitarian and often monumental railroad engineering structures.

Richard J. Cook

Chagrin Falls, Ohio
May 30, 1987

A pair of Baltimore & Ohio diesels roll a coal extra east across the Allegheny River bridge at East Mosgrove. Huge stone piers support this single track truss bridge, which was part of the old Buffalo, Rochester & Pittsburgh. — RICHARD J. COOK

3028
NEW YORK CENTRAL

Crossing Brandywine Creek at Brandywine, Ohio, in 1951, a Cleveland bound New York Central freight is pulled by No. 3028 a 4-8-2. This structure was on a little-known line of the NYC known as the Lake Erie & Pittsburgh which extended from Marcy, Ohio, (Cleveland) southeast to Brady's Lake. The line perched on a valley's edge, leaped across ravines and gullies, and then made a beeline toward Pennsylvania's steel mills. — RICHARD J. COOK

1

Bridge Types

The story of the railroad bridge, in North America, is also a part of the history of the development of the railroad in the United States, and Canada. As the railway pressed ever westward, the bridge became even more important as an aid in delivering goods and people, and it kept trains on the move. This westward expansion across the continent was reflected in the necessity for continued improvements in bridge design in order to accommodate ever increasing loads, as well as, to cross larger and greater bodies of water and deeper and wider gorges.

The idea of a bridge, basically, is to cross an obstacle, whether it be a body of water or a crack in the earth's crust. Imagine a bridge as a wooden plank crossing a small creek. The plank must be long enough to reach each side of the creek or it will collapse under its own weight, or *dead load*. It must be heavy enough to withstand the weight which is crossing it (the *live load*) or it will sag and perhaps collapse. As the weight moves across the plank it causes the board to bounce. This weight is called the *moving live load*. While this may seem simple, the bridge engineer must consider all of the above forces, plus the energy exerted by a strong wind. This force which sets a span to shaking is called the *wind load*.

The distance a bridge extends between supports is called the *span*. The drawings show the difference between a *simple span* and a *continuous span*. The mid-supports for a span are known as *piers* while the end supports of a bridge are called the *abutments*.

The forces which act on a bridge result in *stresses* in its *parts* or *members*. There are two principle types of stress: *tension* — stretching or pulling apart, and *compression* — squeezing or pushing together. Every member in a bridge must be designed to handle either tension or compression, or both.

The first railroad bridges were simple girder spans built to cross a creek or small river, or they were timber trestles with a lattice work of supports, or wooden truss bridges unable to stand any great weight. More often, though, the early railroad bridges were simple arches built of stone that was native to the area.

STONE ARCH BRIDGE

The stone arch, which had served man with great efficiency for hundreds of years, though expensive to build, and time-consuming in its construction, was, at the same time, the most practical for the early railroads and provided great strength. Many of these early stone structures are still to be found along the rights-of-way of some of our American railroads along the East Coast.

One of the first of this type of bridge, and certainly the oldest rail structure still in use in the United States today, is the single arch Carrollton Viaduct of the Baltimore & Ohio Railroad at Baltimore, Maryland. It was completed in 1829. In 1835, the B&O built the eight-arch Thomas Viaduct at Relay, Maryland, which still supports the main line of the B&O running between Baltimore and Washington, D.C.

In the same year, at Canton, Massachusetts, near Boston, the Boston & Providence completed its Canton Viaduct, an engineering marvel of its time. Other stone arch bridges followed. (See the chapter on Stone Bridges.) Later, railroad arch bridges were built of brick, steel and concrete, instead of the native stone.

The first iron railroad bridge in the United States was built by Richard Osborne at Manayunk, Pennsylvania, in 1845. It soon became evident, however, that cast iron was not strong enough for railroad bridges. The number of iron bridge failures grew monthly. Longer and stronger bridges were required to carry the increased traffic. For example, in 1831, the locomotive *De Witt Clinton* weighed 3.5 tons and traveled at a speed of 20 miles per hour; but by 1880 a passenger locomotive weighed almost 70 tons and traveled at speeds greater than 60 miles per hour.

SUSPENSION

After the 1850's the stronger stone arch bridges became quite popular. A suspension bridge at Niagara Falls, however, made a successful debut. (See chapter on John Roebling and His Great Suspension Bridge). In the makeup of a bridge, the greatest strides in bridge building came with the introduction of steel.

Leading Types of Railroad Bridges

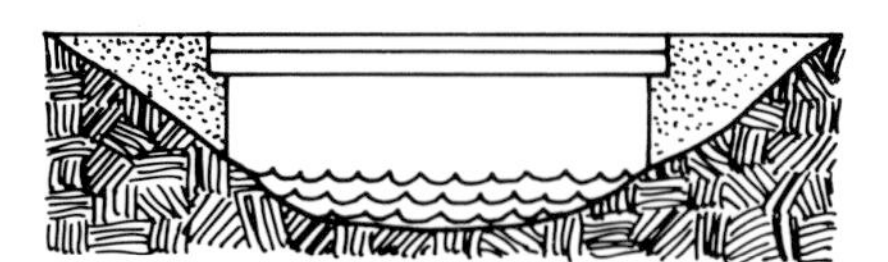

Beam or Girder

Truss

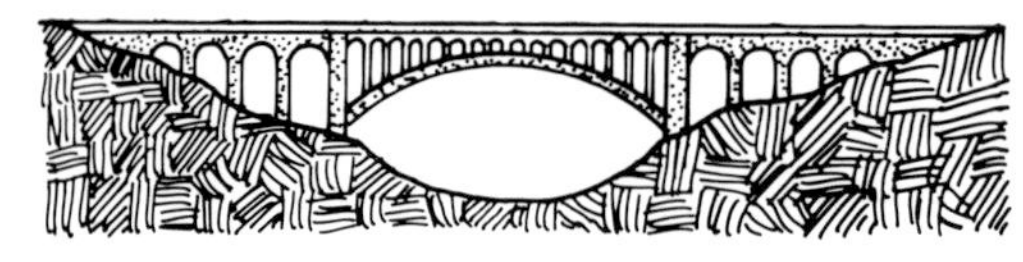

Arch

Reinforced Concrete Arch Viaduct

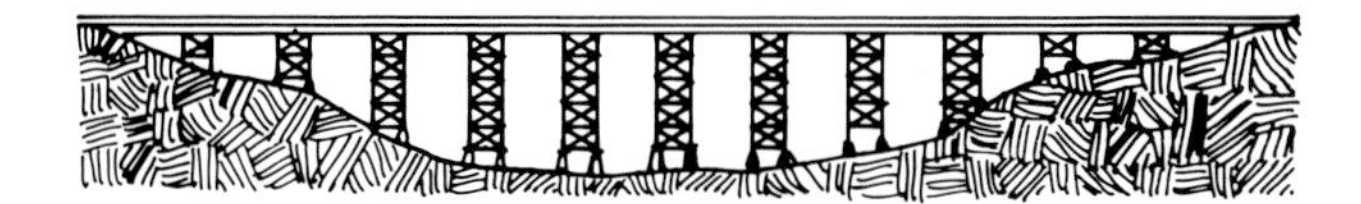

Steel Trestle

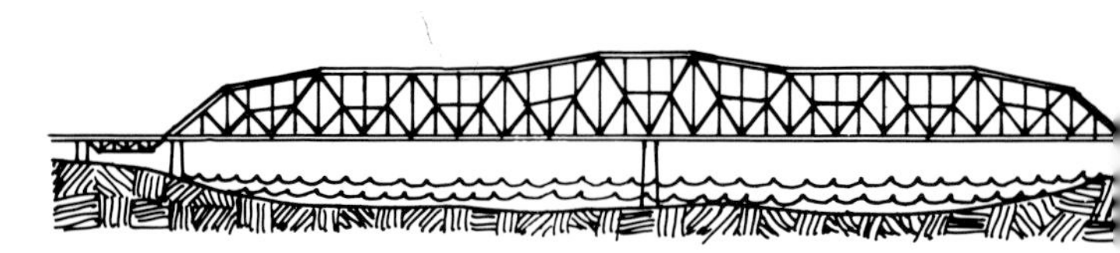

Continuous Truss

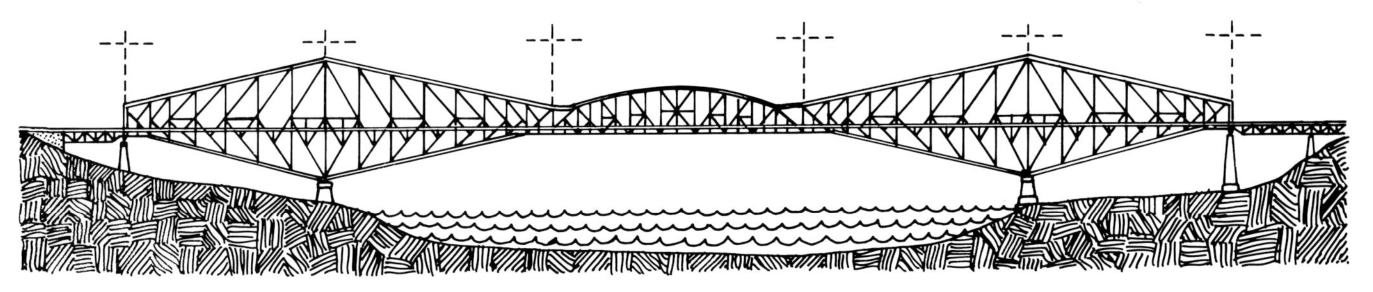

Cantilever (Based on Quebec Bridge)

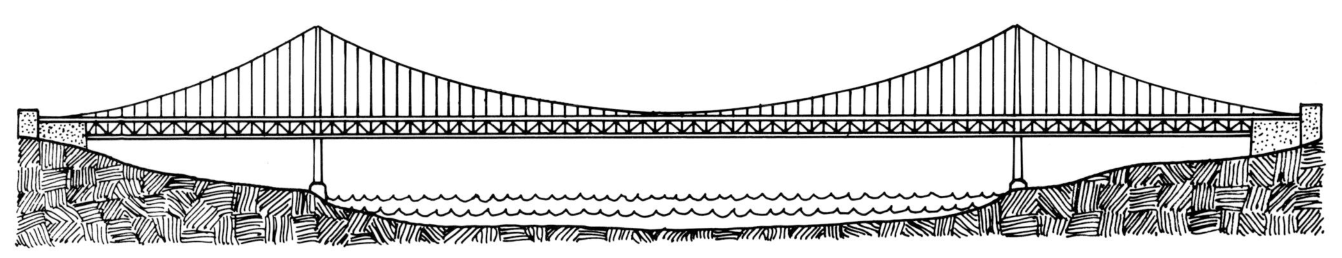

Suspension

As weights of engines and trains grew, the suspension bridge did not turn out to be practical for railroads. This type of bridge is generally too flexible to resist the severe vibrations set up by fast and heavy trains (moving the *live load*). In order to carry the weight of the faster moving trains of the late 19th and 20th centuries, the bridge designers turned to the steel truss.

THE TRUSS

With the perfection of the steel-making process and the availability of metal for bridges, a new era was opened for railroads — and just in time, too! Brittle iron truss bridges, which limited the weight of locomotives and trains, were phased out. Now the trusses could be constructed of steel. The truss, which is probably the most common type of larger railroad bridge built in North America today, is a structure made up of individual members arranged in the form of triangles. Truss spans are either *simple* spans (a support at either end) or *continuous* spans (spanning three or more supports).

CANTILEVER

Truss spans in use on railways can also be of the cantilever design. A cantilever consists of, 1) an *anchor span* between an end support and a tower, 2) a *cantilever arm* that projects beyond the tower and, 3) a *suspended span* that joins the cantilever arm projecting from the opposite direction.

The above types of trusses are all fixed bridges, but the railways also use movable trusses. These movable bridges are built over navigable waterways on which a variety of boats travel, including small pleasure craft and large ocean-going vessels. It is essential that both forms of transportation cross each other with as little interference as possible.

Because of the cumbersome structures required to provide mobility, movable bridges are generally less graceful in appearance than fixed ones. However, the spectacle of one of these great bridges in motion is very impressive.

STEEL GIRDER BRIDGES

For railroad use, long steel girders with an I-shaped cross-section are laid horizontally on piers and the track is supported by cross-beams between the main girders.

The girders may be simple-span (between two adjacent supports) or they may be continuous-spans (over several supports). The structure is called a *rigid frame* when the girders are riveted or welded to support legs so that they form an integrated structural unit. Its rigidity can be identified by the heavy connections required at the corner which is formed by the girder and leg. Long-span girder bridges are constructed of large plate sections riveted or welded together. Preformed concrete girders are now often used for the shorter bridges, supported by pre-stressed concrete piers.

TRESTLE

A trestle is a series of bridges supported on piers. The term trestle is used primarily when spans are short and the piers are the major part of the structure. Viaduct is often used as an alternate word for trestle. Timber once dominated the trestle scene on North American railroads. A wood trestle was faster to erect and cost much less than steel. During the early construction years wood trestles were used as temporary bridges and later replaced by filling in with dirt or replaced by a steel structure. Timber trestles are supported either by bents made of piles driven into the ground, a *pile trestle,* or by bents made of framed posts resting on a foundation, a *framed trestle.* Today concrete trestles are now coming into favor and are used for new construction or for a replacement trestle.

SWINGING DRAWBRIDGE

A swinging drawbridge pivots on a turntable which is mounted atop a pier in the middle of a river, thus allowing boats to pass. The arms are made of cantilevered steel trusses or girders, and are usually equal in length so that they counterbalance. For stability and support, relatively large piers are needed on these turntables. Swing bridges have two disadvantages: They are slow to open and their piers obstruct the navigational channel. However, they do provide full vertical clearance.

The longest of the swing bridges (railroad) is that of the Santa Fe Railway at Fort Madison, Iowa. Its span over the Mississippi River is 525 feet long and was built in 1927.

BASCULE BRIDGE

To overcome the channel-blocking feature of the swing bridge, the bascule principal was developed whereby a truss is hinged at one end while the other end is being lifted into the air. This is known as a *single-leaf* bascule bridge, or "jacknife," and is a common form of a movable railroad bridge. The other bascule type is the *double-leaf* bascule in which two single leaves meet at the center to form, in effect, one bridge over a river. As a railroad bridge, this bascule type is extremely rare.

VERTICAL LIFT

A more recent type of movable bridge, which has seen greater use, is the *vertical lift.* This bridge has two high towers, located on either side of the channel, and the gap between the towers is spanned by a simple bridge, usually a truss which can be moved up and down on the towers like an elevator. The counterweights are located in the towers. The vertical lift bridge can be built more economically than other types of movable bridges. It uses little power and can be lifted at a rate of 50 feet per minute, which is relatively rapid. However, it does not provide a clear opening for navigation travelling below because the bottom of the elevated span limits the height

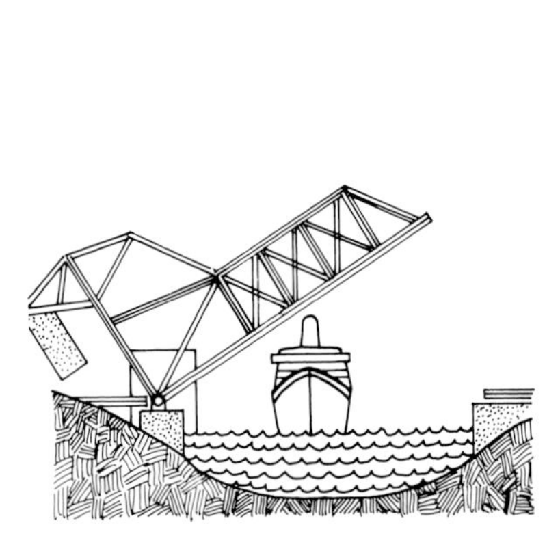

Bascule

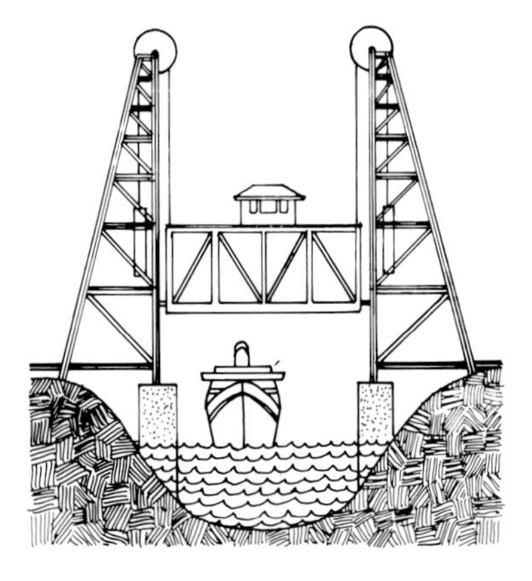

Vertical Lift

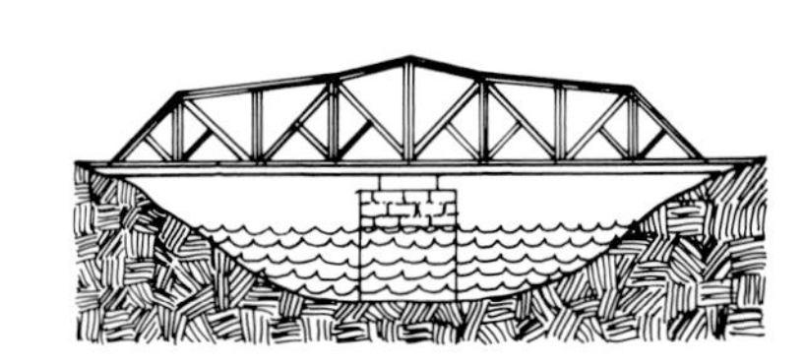

Swing

of ships which can pass underneath.

The longest vertical lift bridges are the Baltimore & Ohio's Arthur Kill Railway bridge at Elizabeth, New Jersey, 558 feet, built in 1949; and the former New York, New Haven & Hartford (now Conrail) bridge over the Cape Cod Canal at Buzzard's Bay, Massachusetts, 544 feet, built in 1935. Both of these bridges are illustrated elsewhere in this volume.

Modern Simple Trusses

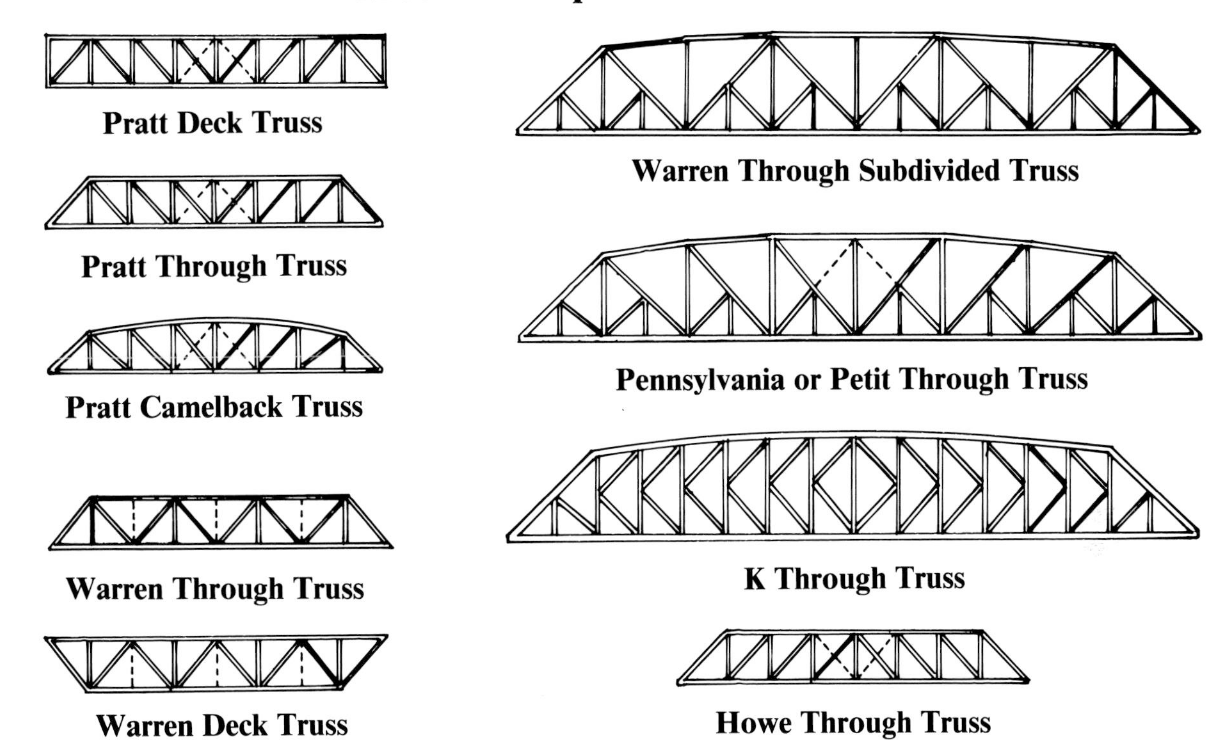

Pratt Trusses

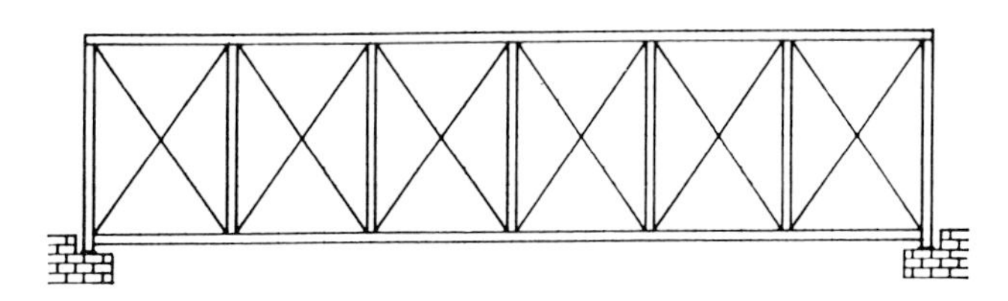

The First Truss
Designed by Thomas Pratt (1842)

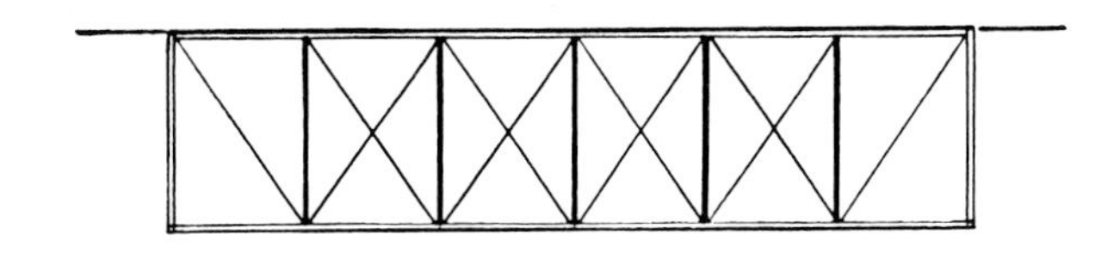

Pratt Deck Truss

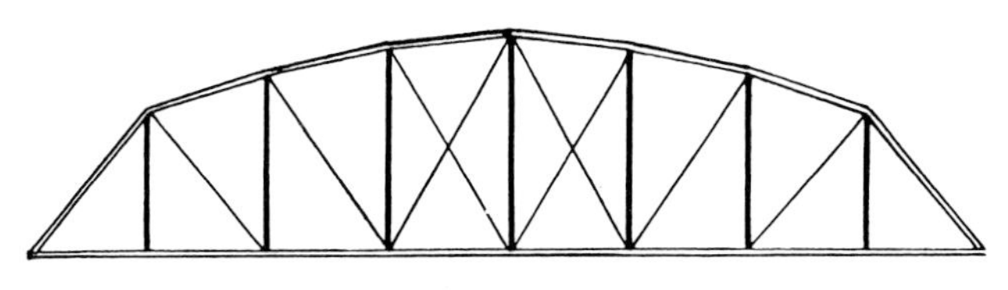

Pratt Through Truss

The Thomas Viaduct became the Baltimore & Ohio's most famous landmark after its completion in 1835. Its best known angle is shown here looking south from Relay as the last true B&O passenger train heads for Baltimore from Chicago on the morning of May 1, 1971. This was a ceremonial pre-Amtrak run of the famous *Capitol Limited,* with two dome cars in the center of the train. — H. H. HARWOOD, JR.

2

Monuments in Stone

Stone, as a building material for bridges, has been in use for centuries. The arch construction, combined with stone's weight and lasting properties, resulted in a structure that was capable of sustaining great weights and also of enduring the ravages of time.

The Pont du Gard near Nimes, France, was built by the Romans during the reign of Caesar Agrippa in 19 B.C. It still stands as a monument to the early Roman engineers who developed and became skillful with this form of bridge building. The earliest specimen of an arch still in existence is in Rome. Built between the sixth and fourth century B.C., it is located over a drain in front of the Temple of Saturn. Great stone arch bridges followed: Le Pont d'Avignon in France, the 103-ft. span of the Ponte San Martino (first century B.C.), various bridges across the Tiber in Rome, the Pontevecchio (1177) in Florence, famous London Bridge built in the 13th century and many others.

Bridge building was well established when the locomotive, which ran on rails, appeared on the scene. The dynamic force that a moving train exerted upon a bridge structure grew as the years produced larger and heavier trains. Thus, for the railroad developer, the decision to build early bridges of stone was a wise one. Whether it was a single arched culvert over a small creek or a many-arched structure in the earliest bridges, in most cases, stone was the material used. When economy took precedence and girder bridges were built of timbers, it was found that it was very necessary to replace and continually strengthen them. The European tradition of stone bridge building carried on to the New World.

On Independence Day, 1828, the merchants of Baltimore launched an enterprise that would bring prosperity to their port city: the Baltimore & Ohio Railroad. The first spadeful of earth for the stone sills that would support the wooden stringers, on which strap iron rails rested, was turned by Charles Carroll. This 91-year old gentleman was the sole surviving signer of the Declaration of Independence.

Therefore, it was quite fitting that when the

The Carrollton Viaduct, built in 1829, is probably the oldest rail structure in the United States still in use today. A Baltimore & Ohio freight, with three diesel switchers, rolls across this double track bridge that is 312 feet in length. The small arch at the far left once crossed a wagon road, and was still visible in 1975 when this photograph was taken.
— H. H. HARWOOD, JR.

road's new stone arch bridge was completed, it be named for the aged patriot. Carrollton Viaduct (for Charles Carroll of Carrollton), completed in 1829, still supports a double-track line of the B&O. It is a single Roman arch 312 feet long, 51 feet above Gwynn Falls, a minor creek, with a 100-foot span.

The structure, which contains 12,000 perches* of stone, is divided into a main arch and a side arch. Originally, the side arch spanned a wagon road. Many tons of the heavy granite were laid upon its temporary wooden centers before it was finally keyed and capable of sustaining itself. The records of construction indicate that the wooden structure held 1,500 tons before the arch was completed. and yet the centers themselves had not yielded so much as one-eighth of an inch!

The stone for the building of the Carrollton Viaduct came, primarily, from the vicinity of Ellicott's Mills. The rest of the stone was quarried at Port Deposit on the lower Susquehanna River.

One of the greatest stone monuments ever built on the B&O system came just a few years later and it, too, was a bridge. Benjamin Henry Latrobe II, son of the celebrated Greek revival architect of Washington, D.C., was the assistant engineer for the B&O at the time. In 1835 he designed and built a most ambitious curving stone-arch viaduct over the Patapsco River at Relay, Maryland, nine miles southwest of Baltimore on the Washington Branch of the new railroad. It was a project which had no precedent in the United States.

The 612-foot structure, laid on a curving alignment, has eight elliptical arches, each 60 feet in width, and carries a double-track about 65 feet above the stream. This viaduct, originally built for the passage of six-ton locomotives, is still in use today and now accommodates 300-ton locomotives with the heaviest of trains. In fact, engineers have determined that the viaduct is so sound and so solid that it carries no bridge rating whatsoever. All of this without any alterations or repairs to the original structure, save periodic pointing of the masonry fabric.

(* a cubic measure for stone, usually equal to 24¾ cubic feet)

The Thomas Viaduct was the first multi-span masonry bridge built for American railroad use and the first in the United States to be laid on a curving alignment. For a time it was known as "Latrobe's Folly" by those who doubted that such a structure could ever be built — or could stand under its own weight. It was named for the road's first president, Philip E. Thomas. Thomas was responsible for forming both the dream and the tangible shape of the Baltimore & Ohio. The viaduct is a fitting memorial to those accomplishments.

> The unique feature of the structure is the placement of the lateral pier faces, which are laid out on radial lines because of the curving plan, each pier thus having a wedge-shaped horizontal section. The bridge was built for locomotives that weighed little more than the 3½ tons of the B&O's 'York', but because of the mass of material in the arches and piers is great enough to absorb the enormous compressive loads and bending forces exerted by modern diesel locomotives. It continues to carry successfully the passenger and freight traffic of the railroad's eastern (Washington-Baltimore) main line. The Thomas Viaduct is a superb work of architecture as well as of engineering: the simple moldings and the rough-faced granite blocks carefully set in uniform joints give the massive span great power and dignity.[1]

In New England another great stone bridge was completed in 1835: the Canton Viaduct on the Boston & Providence Railroad over the east branch of the Neponset River at Canton, Massachusetts. The choice of the route over the Neponset was made at the insistence of Joseph Warren Revere, one of the six members

---

[1] *American Building, Materials and Techniques from the Beginning of the Colonial Settlements to the Present* by Carl W. Condit, University of Chicago Press, 1968.

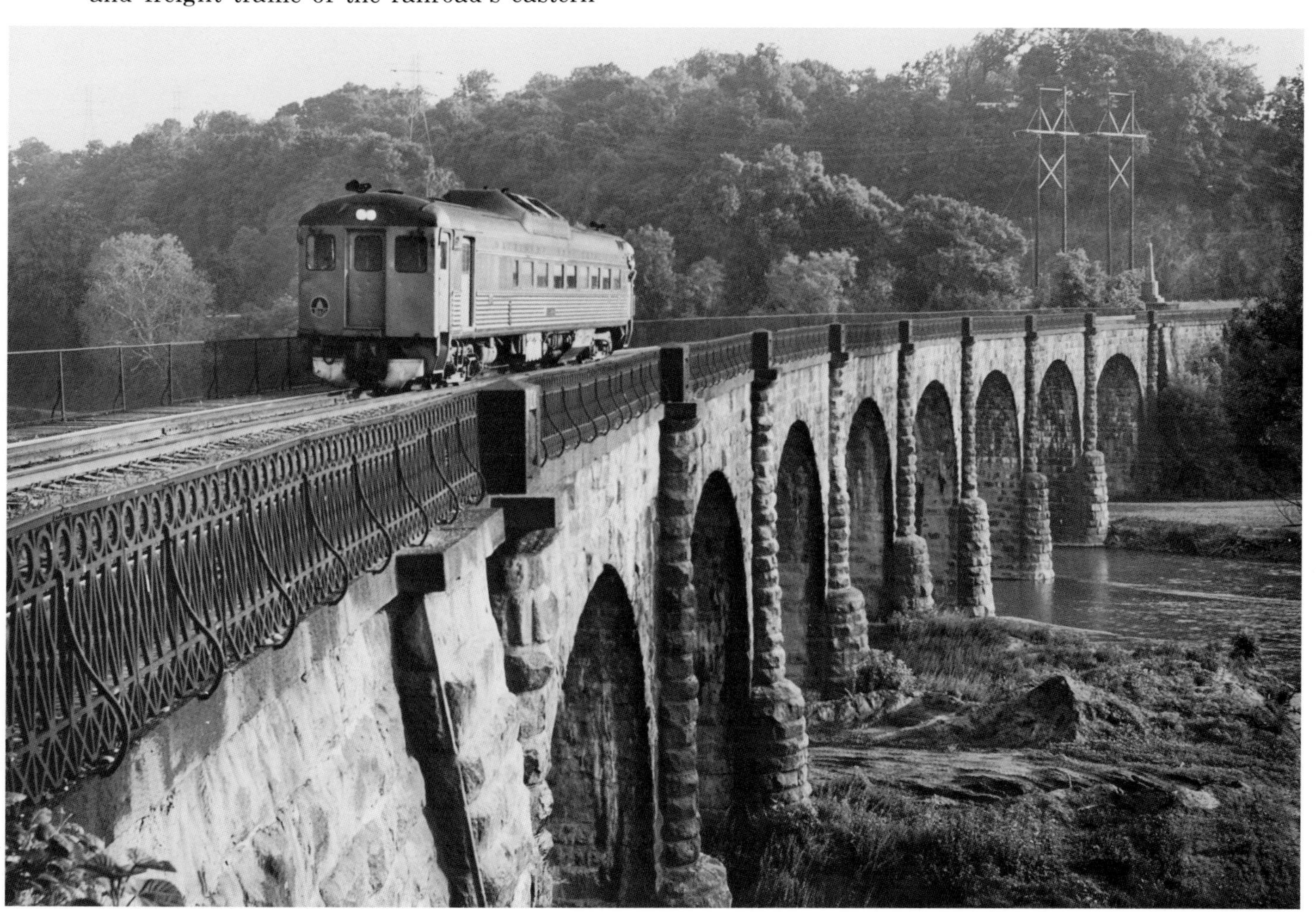

The curved Thomas Viaduct curls across the Patapsco River at Relay, Maryland, just west of Baltimore. A Budd built RDC-2 handled local trains in 1951. — H. H. HARWOOD, JR.

Massachusetts' Canton Viaduct today carries Amtrak trains, those of the Massachusetts Bay Transit Authority, and Conrail freight trains. The stone bridge, built by the Boston & Providence Railroad, in 1835, crosses the east branch of the Neponset River, at Canton, Massachusetts. (ABOVE) A MBTA commuter train heads west. (OPPOSITE PAGE) The west side of a portion of the viaduct overlooks a pleasant waterfall. This close-up photograph illustrates the stonemason's art. — BOTH RICHARD J. COOK

of the board of directors. Revere was the son of Paul Revere who, in later years, had moved to Canton and established a copper and brass foundry on the site of a Revolutionary powder mill.

The viaduct, spanning what was then called the Canton River, was designed by Captain William Gibbs McNeill and Major George Washington Whistler. Both men had worked together on the Baltimore & Ohio and later surveyed and built most of the early rail lines in New England. It is interesting to note that the two men were related by marriage. McNeill's sister, Anna, had married Whistler. Their son, James Abbott McNeill Whistler, born during the construction of the Canton Viaduct, later emerged as a famous painter, best known, perhaps, for painting, *Whistler's Mother,* for which Anna posed 40 years later.

The viaduct proved to be the final link in the new railroad between Boston and Providence. It was built with granite from a local quarry and brought to the site by horse and wagon. The foundations of the viaduct are some eight feet below the surface of the ground. The majority of the interior is hollow but the main

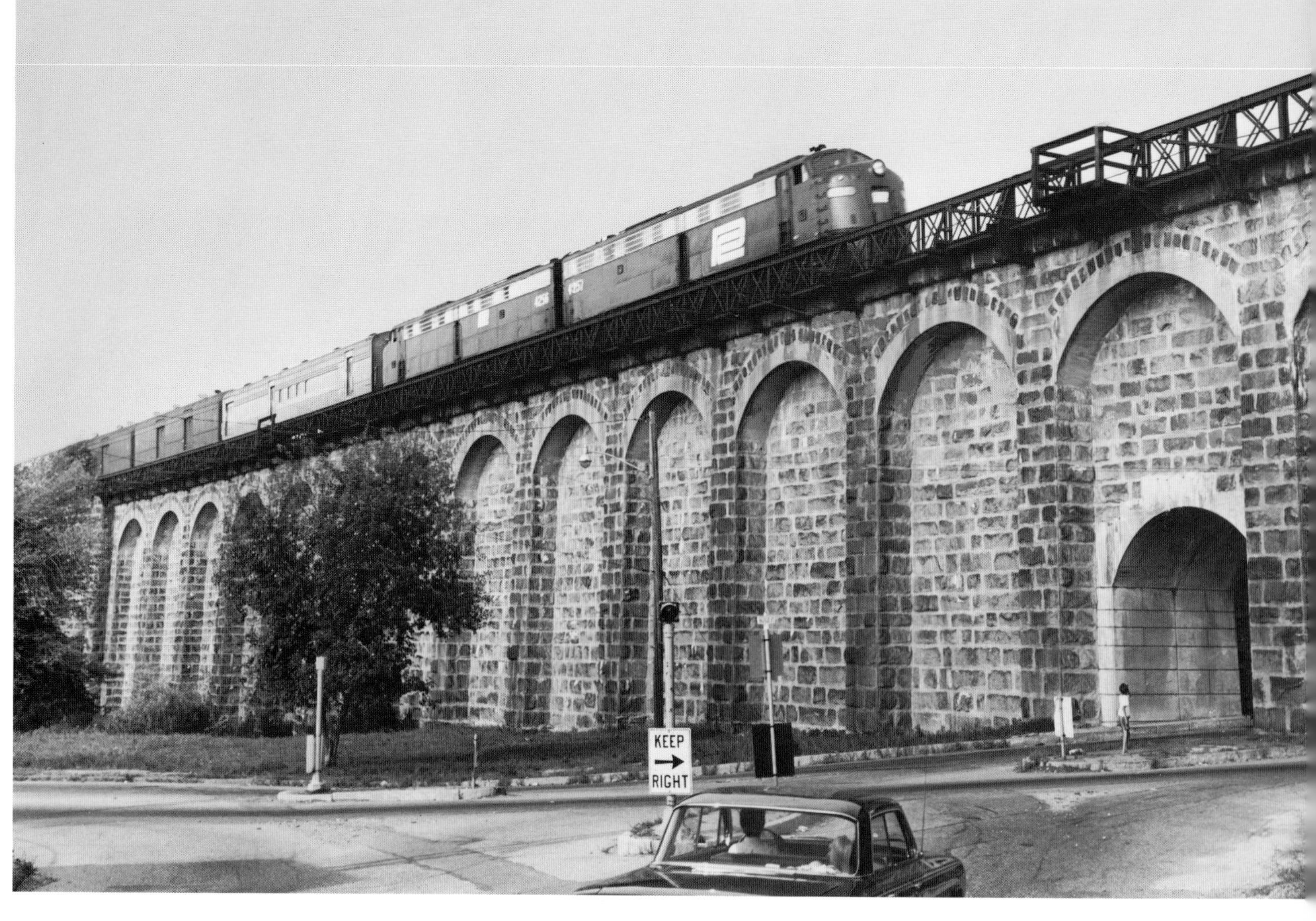

A Penn-Central New York-bound passenger train crosses the Canton Viaduct at speed. This photograph shows how a highway was cut through the original masonry walls. The Neponset River is at the left of the highway. — H. H. HARWOOD, JR.

walls are four feet thick. The arches are filled and are an integral part of the structure, not filled in later, as it might at first appear. The exterior has buttresses connected at the top by segmented arches with the top of the viaduct being finished the full width over the buttresses. The bridge was originally planned for a single set of tracks.

Stretching 615 feet in length, three feet longer than the Thomas Viaduct built the same year, the "Stone Bridge," as it was known in the early days, has a width of 22 feet and is 70 feet above the surface of the river. The original road passed through the structure in a semi-circular arch of 22 feet. Down at the river level there are six semi-circular arches of eight feet four inches each which allow water to pass through the viaduct.

On June 6, 1835, when the viaduct was nearing completion, an article about the structure appeared in the *Providence Journal*. It said:

> ...The viaduct at Canton, though yet unfinished, is a stupendous work. A view of it many times repays the trouble of passing round. The excavation and embankments in Canton are also worthy of minute attention; they testify in strong language man's dominion over nature, and his ability to overcome any obstacle to any understanding that is not morally or physically absurd. The project of cutting through these rocky heights and crossing the valley of the river by the viaduct was a very bold one. A hesitating mind would have surmounted this by a stationary engine or some less formidable way. But any other would have detracted very much from the facilities which give

value to such a road. The road has been constructed under the direction of Major McNeill, and it will stand for ages an enduring monument of the high talents and high attainments of its accomplished engineer.

Five weeks later the viaduct was completed. On Tuesday, July 28, 1835, the first regular train passed over the structure. Traffic on the road grew and the increased volume demanded double-tracking the line, including the viaduct some 25 years later. It was a tribute to the foresight of the bridge's engineers that this could be accomplished with such relative ease.

In 1910 the railroad crossing the bridge, then known as the New York, New Haven & Hartford Railroad, reinforced the arches with concrete on both sides of the structure for additional support.

In 1912 George T. Sampson, one of the railroad's engineers, had one of the larger stones near ground level removed so he could examine the interior of the viaduct. Sampson was joined by Robert Rogers of Canton, a great-great grandson of Paul Revere, and a local historian, who reported in his diary:

> Previous to our arrival three platforms had been erected inside the structure which were connected by ladders. We crawled through the opening and, supplied with electric flashlights and a camera, we had no difficulties as we proceeded with the examination, which proved satisfactory. There were no broken or cracked stones and the joints were still full of good mortar, all of which I think speaks well for these old-time engineers and masons when you consider that this bridge had been built for almost eighty years, and for one of the first railroads that was operated in this country, before they had a chance to learn by experience what would be the effect of the rapidly moving trains or the effect of the constantly increasing loads it has been called upon to carry. For you must remember that when this structure was built it was built to carry a single track and a train with an engine weighing perhaps thirteen tons. That would appear today to be a toy locomotive and be but little larger than the old-fashioned stage coach, which today the engine will weigh a hundred tons or more and many of the loaded freight cars will weigh from seventy to one hundred tons.

The bridge still handles main line traffic on the route of the former New Haven line between Boston and Providence. It was absorbed into the Penn Central System, and now forms a part of Amtrak.

The east side of the Canton Viaduct, unseen by most, blends in with the surrounding countryside making a sylvan setting of peaceful waters and ancient stone. Obviously the Neponset River never carried a large volume of water. — RICHARD J. COOK

Massively built of great stone blocks, the Starrucca Viaduct of the Erie Railroad has met the challenges of time — heavier axle loadings, and higher train speeds. The height of the structure is something often overlooked in photographs of trains. The 100-foot columns seem to dwarf the *Erie Limited* as it crosses the Starrucca and approaches Susquehanna. Today, Conrail trains use this structure.

Rolling hills and woodlands remind one of the grades involved in the area surrounding the Starrucca Viaduct. In this scene helper engines cross the viaduct on their return to Susquehanna. In the foreground, the Delaware & Hudson passes under the viaduct en route to a junction with Erie's coal branches further south. — RICHARD J. COOK

As the nation expanded westward the railroads reached out beyond Baltimore, Boston and New York to the Great Lakes and Chicago. The railroads that got there first were able to follow the banks of rivers to achieve an easy grade. But late-comers, like the New York & Erie Railroad, in order to reach the Great Lakes, had to thrust their lines across mountains and over wide valleys.

Benjamin Wright, surveying the Erie route in 1834, had a goal in mind, which was not to exceed a maximum grade of one percent (one hundred feet per mile). This would be a gentler grade than that of the Baltimore & Ohio Railroad. However, in order to reach the low-level grades and follow the Delaware River Valley from Port Jervis, the railroad had to cross the Shawangunk Mountains in eastern New York State. This required an excavation through the rocky summit of about a half mile long and as much as 50 feet deep. Another cut halfway down the mountain was three-quarters of a mile long and 40 feet deep. After leaving the easy grades of the Delaware River Valley, the railroad builders encountered the deep gorge of Starrucca Creek. Here at Lanesboro, Pennsylvania, they decided to build a 1,200-foot stone arch viaduct to bridge the gap.

Financed with British funds, the railroad apparently felt that it could afford to embark upon a project that, for its time, turned out to be the most expensive bridge in the world. Thus in 1847, Julius W. Adams, superintending engineer for the railroad's construction, enlisted James P. Kirkwood as engineer to aid in this project. When completed in 1848, the 18-arched viaduct stood 110 feet above

Starrucca Creek. It had cost $320,000 in materials and labor.

The bridge is a masterpiece of engineering beauty. It has been in continual use for over 130 years and has required little maintenance. Although designed to carry 50-ton locomotives, it has since handled the weight of some of the heaviest steam and diesel locomotives ever built.

One of the last great stone viaducts, the Starrucca, was built as an original structure and not as a replacement for a previous bridge. It is interesting to note that the base portions of the piers and the covering of the deck are of concrete, perhaps the first structural use of concrete in American railroad bridge construction.

The following is an account by Hosea M. Benson of Jackson, Pennsylvania, published by the Montrose (Pa.) *Independent* December 24, 1931. Benson was born on July 1, 1837 and died in 1937 at the age of 99. He witnessed the construction of the Starrucca Viaduct:

> There was great commotion among the people when the first report came that there was to be a railroad built from New York to Lake Erie, there to connect with boats to go to the far west. It was in the year 1845. It was some time before the surveyors got up to Gulf Summit Cascade and to Lanesboro. Here they found the most expensive seven miles of railroad to build in the United States, as reported at that time. The viaduct at Lanesboro had to be built across the Starrucca Valley 1200 feet, 110 feet high — requiring nineteen piers and eighteen arches.
>
> My father and I visited the scenes and work of the railroad building very often, which was very interesting and exciting to us all. I have never forgotten how they built it.
>
> Two men, Barker and Denton, had the contract to build the bridge for $375,000, as reported, but it was said that they lost out and went bankrupt. They first dug the foundation pits to bedrock. These pits were deep, as deep wells. They were dug larger than the pier. All this work was done by handpower — pick and shovel and windlass.
>
> I remember well seeing four men at a plunger pump to keep water out. When the pits were dug, they were filled with pounded stone and cement to even with the surface, and foundations were as smooth and white and nice across the valley ready for the piers.
>
> The company gave work to all the people they could. To fill those foundation pits with stone and cement was a big job. They drew the stone from a quarry near Brandt with all kinds of teams of horses and oxen and wagons.
>
> It was said the twenty feet or more of the foundations in the ground cost more to build than any other twenty foot section of the piers.
>
> Now with the foundations ready comes the building of the piers. They built what they called falsework between the piers of timbers, scored and hewed about 16 feet high, and laid a wood track over the piers. At the north end they ran the track on the bank and up the creek near Brandt.
>
> The stone was cut and numbered and loaded on the stone cars drawn by horses and mules over the piers and were unloaded by derricks down on the piers. They drilled two holes in the large stone about two feet apart. They had a short chain with ring in center and shortplugs on each end. They would stick these plugs in the holes and let them down on the piers where they were fitted to go. The stone was cut, numbered and marked with black paint. The masons on the piers knew where to lay them.
>
> When the piers were up to the track, they built another section of falsework and so continued to raise the track over the piers until the work was done. When the piers were high enough for the arches, they left projecting a row of stone to set the wood arches to lay the stone on. They were now 100 feet from the ground and every pier was fastened just the same and stands in a perfect row. Just the same as was built. It is a wonder how they got such large top stones on top of the bridge. They lay today just as they were laid 85 years ago.
>
> The bridge was built for those light wood-burner engines and light trains but it now bears the burden of the Matt Shay engines, long and heavy trains. The bridge is one of the world wonders.
>
> The Erie Railroad Company ought to erect a monument to the memory of Barker and Denton, contractors and builders of the Starrucca Viaduct, 1845-1848.
>
> When the bridge was completed it was a wonderful view to behold, to see that bridge with the falsework of timbers filling the space between the piers from the ground to the arches, 110 feet from the ground.

Among the many large bridges that form a part of the old Philadelphia & Reading Railway, there are probably none better known than those that cross the Schuylkill River at the Falls of Schuylkill, at Philadelphia, and at Peacock's Lock near Tuckerton, five miles above Reading.

When the line was extended from Reading to Mount Carbon, the bridge at Peacock's Lock was constructed of wood and was designed to carry only a single track. Known as a Town lattice truss bridge, it was 675 feet in length and was erected during the year 1839. Apparently the plans were prepared by Richard B. Osborne, who directly supervised its construction. Osborne, at that time, was an assistant engineer but later became chief engineer of the company. He has also been recognized as constructing the first metal railroad bridge in the United States, the iron truss built at Manayunk, Philadelphia, in 1845.

During the first few years of operation, the road equipment used was so light that the

Reading Railroad RDC cars cross the Peacock's Lock bridge at Tuckerton, Pennsylvania, as they roll eastbound from Pottsville to Reading in August 1972. (BELOW) The Peacock's Lock bridge is quite unusual because of the large circular openings in the spandrels. — BOTH H. H. HARWOOD, JR.

In 1855 the Reading Railroad's Falls Bridge was completed and opened for traffic. This seven-span stone structure is located at the falls of the Schuylkill River in Philadelphia. Wayne P. Ellis photographed a steam passenger train on the bridge before the diesel era. — H. H. HARWOOD COLLECTION

wooden bridges were more than sufficient to carry their weight. Gradually, as heavier equipment was placed in service, the wooden bridges had to be strengthened. By the year 1847, additional arches, piers and supports were required to make the Peacock's Lock bridge safe for the passage of the 38-ton locomotives then coming into use.

The first wooden bridge built at the Falls of the Schuylkill in 1839 was destroyed by fire on the night of August 25, 1842. It was immediately rebuilt of wood, to a length of 615 feet and of the same general design as the one at Peacock's Lock, near Tuckerton. The only exception was that a portion of the bridge was built of truss work supported by piers and intermediate trusses.

In 1848, G. A. Nicholls, engineer and general superintendent, recommended that the wooden structures at the Falls and at Peacock's Lock be replaced by stone arch bridges of the designs that he had prepared.

In their 1852 fiscal year report, the president and managers also recommended that authority be given to construct these bridges. Accordingly, at the annual meeting of the stockholders held on January 10, 1853, authority was given to carry out the improvements recommended by management.

The contracting firm appointed to carry out the construction failed, and Nicholls was forced to complete the work with his own forces at a cost much greater than the original estimate. Nicholls, in his report for the year ending November 30, 1853, stated that all the piers of the Peacock's Lock bridge, two abutments and one pier of the Falls bridge had been constructed. In 1853 all the piers of the Falls bridge were completed with the exception of the center pier. In 1855 both bridges were completed to the coping and railing which were added the following year. These bridges were built without stopping road traffic for a single day. Nicholls, in his final report on the work called attention to the fact that the bridges "are of extraordinary strength and durability." It's hardly probable that he had any conception of the weights and stresses that the bridges would undergo in the years to come.

The Peacock's Lock bridge (named for a lock on the Schuylkill Canal) has nine spans of 64 feet each; the distance from the low water mark to the top of the coping is 59 feet. It is unusual in design because of the large circular openings in the spandrels.

The Falls bridge contains seven spans, six of 78 feet each, and one of 36 feet, and the distance from the low water mark to the top of the coping is 48 feet.

Amtrak's Schuylkill Bridge at Philadelphia carries the high-speed "Northeast Corridor" trains across the Schuylkill River. The structure adds a note of graceful antiquity to Philadelphia's famous Fairmount Park. This stone bridge was built by the Pennsylvania Railroad; it then became a part of Penn-Central, and then Conrail before Amtrak took over the New York-Washington route. — RICHARD J. COOK

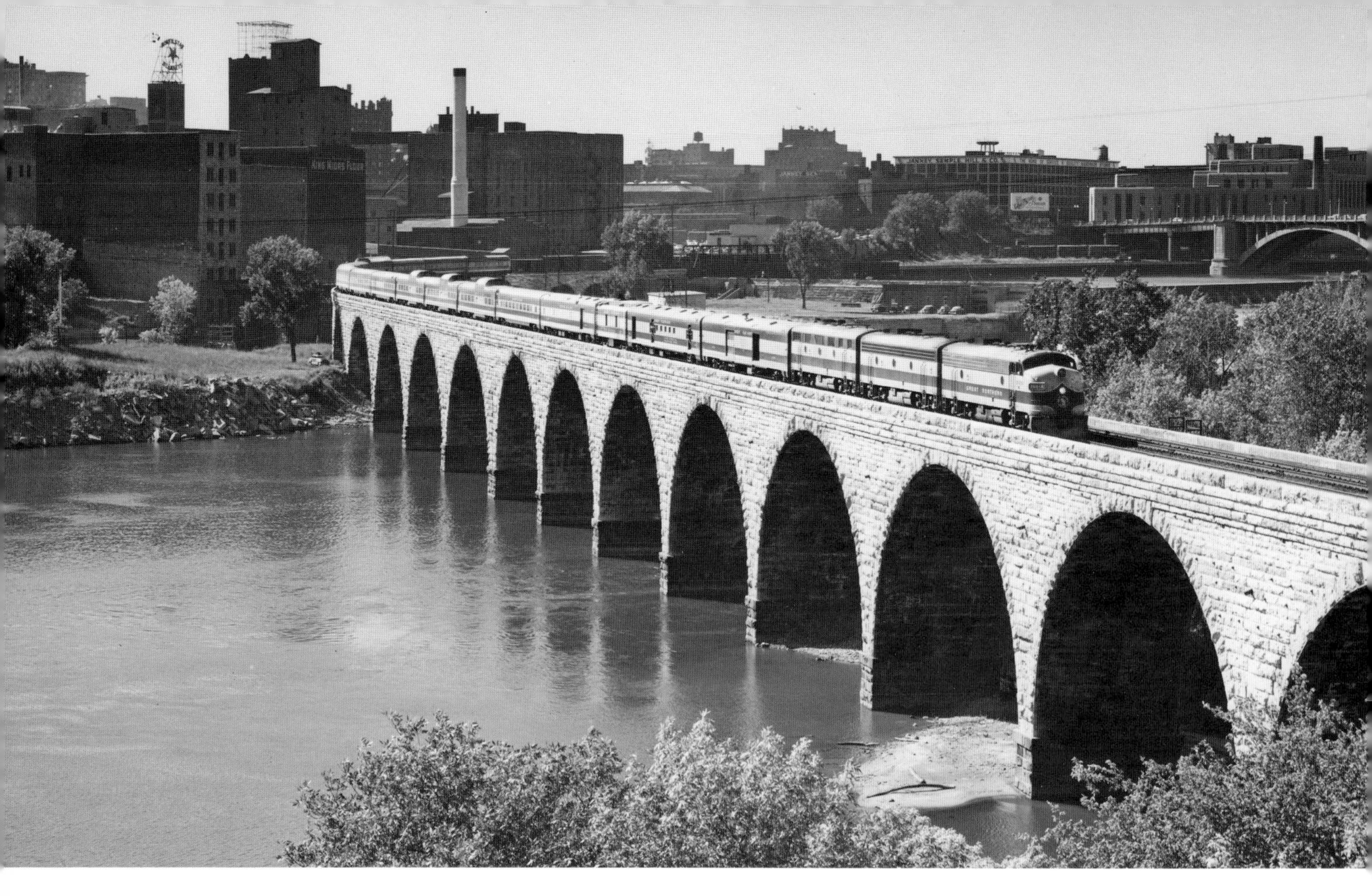

Great Northern Railway's famous streamliner *Empire Builder* is crossing the historic Stone Arch Bridge, spanning the Mississippi River at Minneapolis. A six-degree curve in the bridge was thought impossible at the time and James J. Hill once stated it was "the hardest undertaking I ever had to face." The structure is now part of the Burlington Northern system. — BURLINGTON NORTHERN

"The hardest thing I ever had to do, or the hardest undertaking I ever had to face was the building of the stone arch bridge." That was the statement made by James J. Hill, builder of the Great Northern Railway, reminiscing in 1905 over his brilliant career as the "Empire Builder of the Northwest."

The stone arch bridge, over the Mississippi River at Minneapolis, was one that many competent engineers said could not be built. Just as they had said about the Thomas Viaduct on the Baltimore & Ohio, where it, too, involved a curving route: a six-degree curve. Once again, like the B&O "Latrobe's Folly," Jim Hill's "folly" proved the critics wrong.

The bridge began to take shape in 1882. It was a 2,100 foot long stretch and measured 76 feet high and 26 to 28 feet wide. The 23 arches have diameters of 98, 80 and 40 feet. About 1,283 feet from its east end the bridge is straight, but on the last 817 feet of the west end there is a six-degree curve.

Even though the bridge presented problems never before encountered in bridge engineering, the structure was built. Despite the fact that the rock on which the bridge foundations rested was, in many places, disasterously unstable, the stone arch bridge was completed and opened for traffic in 1883. The concrete, limestone and granite structure had cost $690,000, but, nevertheless, it was to provide the Great Northern and all but three lines which served both St. Paul and Minneapolis with a most efficient crossing of the Mississippi. It has carried the passenger trains of the Great Northern; Chicago, Burlington & Quincy; Chicago & North Western; Chicago Great Western and Northern Pacific.

In 1907 Great Northern bridge crews started a four-year reinforcing program, placing steel tie rods through the structure, installing new drain pipes and placing reinforced concrete arch rings about the stone arch rings. In 1925, with locomotives increasing in size, the two tracks were moved six inches farther apart and some of the stone coping was trimmed off so that locomotives and cars would clear it without difficulty.

In 1962, after more than 15 years of talking and planning, the U.S. Army Corps of Engineers decided to go ahead with its St. Anthony Falls Project of extending river barge navigation 4.6 miles upstream, deepening the channel and replacing arches Nos. 13 and 14 with a 200-foot inverted Warren steel truss span. This project became necessary in order to provide a 56-foot clear channel for a lock entrance below.

In 1965 another important event occurred in the life of the Stone Arch Bridge, as it is known on the railroad. the publication, *Railway Track and Structures* for September, 1965, reported as follows:

> Being called out at night on an emergency is not new to most railroad engineers, and often the problems encountered are of a routine nature. Sometimes, however, the problems are extremely serious and may require unusual skill in overcoming them.
>
> Such was the case when George V. Guerin and R.W. Gustafson, Great Northern's chief engineer and bridge engineer, respectively, were summoned from their homes on Sunday evening, April 18, 1965. Something had happened at the road's famous Stone Arch Bridge that demanded their attention.
>
> A switch crew taking a cut of empty equipment from the Minneapolis passenger terminal at about 9:00 p.m. reported it had experienced a sharp jolt while crossing over the bridge. The crew had traced the jolt to a sag in the surface of the structure about one-third of the way across the flood-swollen Mississippi River from the east bank.

Jim Hill's "Folly," now owned by Burlington Northern, has for over a century served as an important trans-Mississippi link in a heavy-duty transcontinental rail system. — RICHARD J. COOK

> This structure had withstood many a rampage of the Mississippi in its 82-year history and neither ice nor high water had disturbed its massive granite-faced piers. Even so, Guerin and Gustafson were concerned. They had been struggling with the problem of high water in the Mississippi for more than a week. The river had risen to the highest level on record and had crested at Minneapolis just two days before. The situation seemed ripe for serious trouble....

What they found *was* a sag — about 100 feet long, centered over Pier 7. Later inspection uncovered the fact that Pier 7 uniformly settled 14 inches with no discernable change in vertical alignment. Several arch stones had

In order to accommodate barge traffic, the Army Corps of Engineers replaced, in 1962, arches Nos. 13 and 14 with a 200-foot Warren truss span. For some reason this famous Stone Arch Bridge looks blemished in its old age. — BURLINGTON NORTHERN

Pennsylvania Railroad's second bridge at Rockville, crossing the Susquehanna River, was this iron truss structure carrying two tracks. Completed in 1877, it had, in turn, replaced a single track bridge of wooden spans built in 1849.

dropped to the river and there was some undercutting of the piers. They were able to correct this by thoroughly grouting the concrete backing of the arches, reinforcing Piers 6, 7 and 8, using cast-in-place Augercast piles, and with the construction of reinforced concrete liners two-and-a-half feet thick on the undersides of Arches 6-7 and 7-8.

The bridge, now the property of the Burlington Northern System, continues to give the same heavy duty service which it has provided for over a century.

The longest stone-arch railroad bridge in the world replaced a wrought-iron truss bridge which, in turn, had replaced a wooden arch bridge, the majority of which was covered.

The Rockville Bridge, as it is known, carried the main line rails of the Pennsylvania Railroad over the Susquehanna River at a point five miles above Harrisburg, Pennsylvania. The present four-tracked bridge, built in 1901, has a total length of 3,820 feet, a minimum width of 52 feet overall and a height of 51 feet from the low water level to the coping. It has 48 70-foot segmental arch spans, which are built with cut-stone voussoirs, ashlar pier faces and spandrel walls and concrete hearting and spandrel filling.

The present four-track stone arch bridge, finished in 1901, is the largest of its type in the world. Locomotive No. 350 takes the first westbound train across the new structure. The iron truss bridge may be seen at the left.
— W. H. JENNINGS

A 1917 view of the Rockville Arch Bridge shows a steam powered passenger train crossing this world famous structure. The west bank, shown in the foreground, is lined with summer fishing cabins.

It contains about 100,000 cubic yards of masonry of all sorts. The increase in volume of traffic, weight, dimensions of rolling stock, and improvements of roadbed made it necessary to build this bridge as a replacement of the double-track, through truss iron bridge of 23 120-foot spans, which had been built in 1874 near the same site.

Now under the ownership of Conrail, the Rockville Bridge was recently designated as a National Historical Civil Engineering Landmark by the American Society of Civil Engineers. It has withstood many bad floods and, also, Hurricane Agnes, of 1972, which proved to be the most destructive thing that has ever hit the Susquehanna Valley.

The Rockville Arch Bridge has varied little in 86 years of service, however, the motive power has changed from steam to diesel. Some years ago, in the above scene, two Penn-Central diesels handled an eastbound freight drag. At the time the Susquehanna River was at near flood stage. — RICHARD J. COOK

Pennsylvania Railroad's Shock's Mill Bridge, 20 miles southeast of Harrisburg, was built in 1904 and contained 28 stone arches in its 2,200 foot span over the Susquehanna River. Today, this bridge connects Conrail's network of main and branch lines with Enola Yard on the west bank of the Susquehanna River near Harrisburg. — HARRY P. ALBRECHT COLLECTION

Another Conrail (former Pennsylvania Railroad) stone bridge of note, crossing the Susquehanna, is the Shocks Mill Bridge, 20 miles southeast of Harrisburg at Shocks Mill, Pennsylvania. Built in 1904, the 2,200-foot viaduct consisted of 28 stone-arch spans averaging 80 feet in length. It carried the electrified freight line that connects Conrail's network of main and branch lines, east of the river, with strategic Enola Yard on the west bank of the Susquehanna, about 24 miles north of the bridge.

Unlike the Rockville Bridge, however, it fell victim to the ravages of Hurricane Agnes. For on July 2, 1972, as a result of the June floods, about one-third of the Shocks Mill Bridge collapsed into the Susquehanna.

A herculean effort to span the gap in this arch structure was mounted by construction forces, at a cost of $4.5 million. It was accomplished with a nine-span double-track, deck-girder bridge supported on eight new concrete piers with an average height of 60 feet.

Two additional stone arch bridge classics. (LEFT) At Waring, Maryland, an Amtrak train crosses the old Baltimore & Ohio's Big Seneca Creek Viaduct on the B&O's Metropolitan branch in 1978. (RIGHT) A Reading Railroad eastbound multiple-unit commuter car rushes across the Wissahicken Creek Viaduct, Philadelphia, the same year. — BOTH H. H. HARWOOD, JR.

Down south, railroad bridges were made of brick rather than stone. This old brick arch bridge of the Central of Georgia Railroad, at Savannah, has seen better days. — RICHARD J. COOK

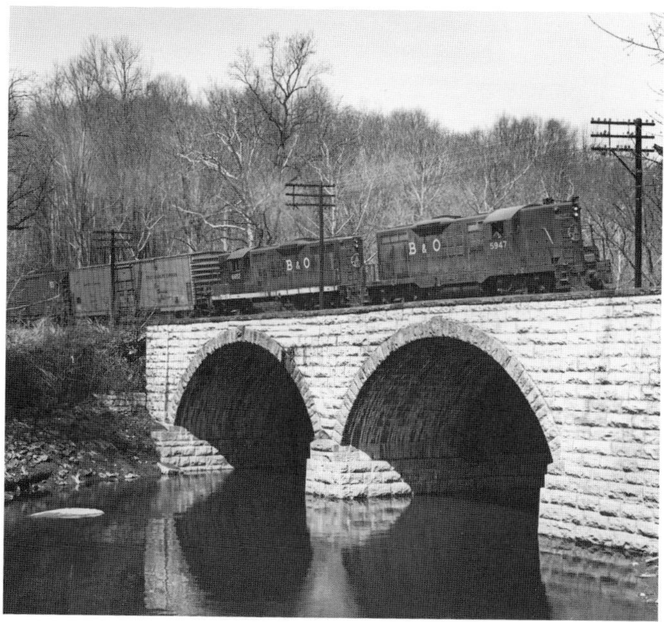

Two Baltimore & Ohio GP-9's wheel a local freight over a graceful two-arch bridge crossing Catoctin Creek at Brunswick, Maryland. — H. H HARWOOD, JR.

One of Pennsylvania Railroad's famous GG-1 electric locomotives glides a New York-Washington train over the Delaware River bridge at Trenton, New Jersey, in 1968. — H. H. HARWOOD, JR.

This Baltimore & Ohio stone arch bridge is over Brandywine Creek at Wilmington, Delaware. B&O's fixation with stone bridges ended by the mid-1830's; afterwards, most new construction was wood, then iron. Leonor Loree's reconstruction of the B&O in the early 1900's briefly brought back masonry bridges, this one circa 1910. — H. H. HARWOOD, JR.

This old stone arch carried the rails of the Allegheny Portage Railroad. At this point, canal boats were placed on flat cars for their trip across Pennsylvania's Allegheny Mountains. The portage line opened for service March 18, 1834. — RICHARD J. COOK

In this rural setting the Pennsylvania Railroad built a stone arch viaduct to cross Conestoga Creek, near Lancaster, Pennsylvania. Here, a Conrail eastbound freight was photographed during December 1977. Note the steel bracing of the arches which enabled them to support the heavier loads of today. — H. H. HARWOOD, JR.

Loree's reconstruction of the Baltimore & Ohio also included a line relocation at Lodi, Ohio. As part of this work, this beautiful stone arch bridge was built in 1907. A date stone may be seen at the left close to the track ballast. In this scene, a freight thundering across the Black River stone bridge, in 1950, was photographed by the author. (RIGHT) The Boston & Maine's venerable stone arch bridge over the Connecticut River at Bellows Falls, Vermont, has had its arches strengthened by means of steel bars bolted through to the other side. Here, in the 1940's, a Boston-bound train is pulled by Pacific-type No. 3660. — JOHN KRAUSE

Stone Arch Bridges on the Pennsylvania Railroad

The Pennsylvania Railroad always maintained their bridges to the highest standards. After all they were to become the "Standard Railroad of the World!" When the construction of the road first commenced, iron railway bridges were almost unknown, and all the bridges on the eastern part of the line were constructed of timber, and, in some cases, for short spans, of stone or brick. This quickly changed as the bridges could not stand the weight of heavy trains. Stone soon replaced wood. The August 13, 1887 issue of the periodical *Engineering News* carried these classic drawings of stone arch bridges on the Pennsylvania Railroad. (TOP) The original Conemaugh Viaduct was destroyed by the Johnstown Flood in 1889. It was replaced by this single arch bridge in the early 1890's. (SECOND) The Pequea Creek bridge, located east of Lancaster at Herrville, was originally a wood truss bridge and this three arch stone bridge replaced it. (THIRD) The three arch stone bridge over Little Juniata creek near Birmingham replaced a part stone and wood truss built in the 1870's. The location of "Big Chiques" bridge is at Landisville, Pennsylvania. Note that former bridgeseats/abutments have been worked into the new structure.

Conemaugh Bridge

Pequea Creek Bridge

Little Juniata Bridge

Big Chiques Bridge

3

It Started With A Kite

The story of a bridge over the chasm of the great gorge of the Niagara River is an integral part of the westward thrust of railroads in the United States. The bridge proved to be an essential cog in the wheel of the accelerating industrial revolution and in the development of America.

The State of New York and the Canadian Province of Ontario are separated by sheer cliffs of 200 feet to the wild chasm below. The boiling water of the Niagara River sweeps away everything in its path. It was at a point about two miles below the falls of the Niagara that in 1815 Major Charles B. Stuart, one of the first promoters of a bridge across the gorge, recommended that a structure be built. Major Stuart was struck with the possibility of connecting, with a "hanging bridge," the Great Western Railway of Canada with what later became the Rochester and Niagara Falls Branch of the New York Central. So he invited the best engineers of America and Europe to contribute their thoughts on the possibility of such an undertaking.

The concept of the suspension bridge was not a new one. Small suspension bridges had been built in the United States, England, and France. But some of them had met with disaster. In England, in 1831, a suspension bridge had collapsed under the feet of troops, all marching in unison. Two hundred soldiers in France had been killed under almost identical circumstances. A bridge at Covington, Kentucky, over the Licking River, had collapsed under droves of cattle. Other such bridges, however, had held and were successful.

There was a stigma attached to the idea of a suspension bidge, especially one across such a wild and dangerous gorge. Most of the experts declared such a scheme — to carry a railroad over the gorge — to be quite impossible. Most of the replies said: "Perish the thought!" and some contained dire predictions as to the fate of such a bridge and to the foolhardy designer who might erect it. Vibrations set up by so heavy a moving load as a train, they declared, would quickly destroy any wire-hung bridge.

Still, the idea, of a railroad linking the Buffalo region with southern Ontario and Detroit beyond, made a great deal of sense to Canadian and American railroad people.

Of all those candidates who were canvassed on the project only four thought they could do the job and only two of these had ever built a suspension bridge; Charles Ellet and John A. Roebling.

The first chance to build the bridge went to Charles Ellet who, in 1845, was the best-known bridge builder in America. He was also the most flamboyant, a tireless actor, and always seeking publicity. He was impetuous and colorful, but also an expert at bridge building.

One of the first problems to be faced at the Niagara gorge was how to get a wire across. Ellet solved the problem by offering a prize of five dollars to the first boy who could fly a kite over to the Canadian side. After many attempts throughout the day, young Homan Walsh accomplished the feat and proudly won his reward. Eighty years later, when he was an old man residing in Lincoln, Nebraska, Walsh's most precious memory was of his boyhood feat — his part in the construction of the first span across the Niagara.

The first strand was spun by means of the kite strings, then larger cords, then hemp rope. Each time, increasing in size and finally the wire cables. After he had one or two cables up, Ellet decided to give vent to his flair for the dramatic and, in so doing, cash in on his exploit. He had an iron basket made big enough to hold him and had it attached to the cable with pulleys. On a morning in March 1848, he stepped inside and pulled himself over the gorge and back again — to the great excitement of crowds on both rims.

"The wind was high and the weather cold," he wrote, "but yet the trip was a very interesting one to me — perched up as I was two hundred and forty feet above the rapids, and viewing from the center of the river one of the sublimest prospects which nature had prepared on this globe of ours."

Ellet was the first man to cross the gorge and he made sure the world knew about it. He apparently thought it necessary to underscore his feat, because several weeks later, after a plank catwalk had been strung, he drove in a small horsedrawn carriage across a swaying span that had no guardrails. Standing up, Ellet drove it like a charioteer. The crowds watched aghast, and women fainted. Not until twelve years later was this daring feat surpassed when Blondin, the famous tight-rope walker, crossed the same gorge by walking on a wire rope, pushing a wheel-barrow and carrying a man on his back.

For ten months, after its completion in July 1848, the temporary span was used as a public footbridge. During this time, Ellet pocketed almost $5,000 in tolls.

The financial supporters of the railroad bridge, however, were becoming increasingly unhappy, since no progress was being made on the main project and Ellet had already stretched himself thin by contracting for a 1,010 foot suspension bridge over the Ohio River at Wheeling, West Virginia. Besides, the Niagara bridge directors wanted the $1.00 tolls that were being collected. The dispute finally ended in Ellet's resignation, his "noble work to stand firm for the ages," as he initially described it, not even started. Ellet's great chance to achieve the acclaim he so desired, praise for an outstanding and beneficial work, was thrown away in a fit of pride.

The Niagara bridge job would wait for a quieter, more careful workman to complete it. That man would be John A. Roebling. Finally, after the project lay dormant for two years, the officials were stirred to renewed activity, and because, originally, Roebling's bid on the job had actually been lower than Ellet's, the directors lived to regret that first decision. During that two years, Roebling, painstaking, methodical, everything that Ellet wasn't, had watched the project with great interest. He was the first to proclaim the possibility of building suspension bridges to carry railroad loading and also, contributed more to a real solution of the problem than any other engineer of his time. So in order to complete their project, it was to Roebling that the directors now logically turned.

Freed from Ellet's price-cutting competition, Roebling undertook the work on a new basis: a fixed salary for his engineering and super-vision, with the owners paying for all labor and materials.

Roebling began his construction in 1851. For four years the work continued, summer and winter — and winters at Niagara can get nasty indeed. But Roebling was a steady worker and he also expected his bridge forces to work steadily. Every step of the erection was planned and closely supervised by him. He had worked his ideas out carefully and was scrupulous in seeing that all details were done with care and were done right.

He was the first to recognize the importance of a heavy stiff roadway. He also utilized the best ideas of previous engineers in England and Europe. His bridge over the Niagara, he had written, would stand up under a moving train because he would make it stiff enough to do so. He designed the two floors of the bridge and the open timber trusswork that was to bind them together as one enormous "hollow straight beam," The timber would be well seasoned, well painted, and the upper floor, where the trains would cross, would be caulked and painted as thoroughly as a ship's deck. As a result, it would serve as a protective shelter for the lower floor and trusswork — similar to an old-fashioned covered bridge.

The *Scientific American* for June 5, 1852 reported:

> A convention of railroad Directors and Bridge Companies was held at Niagara Falls two weeks ago to settle permanently their plans for the construction of the Great Western and Rochester, Lockport and Niagara Falls Roads, and for building immediately a new and greatly enlarged suspension bridge which is to connect these two great thoroughfares. It is not supposed that it is the intention of the companies to build this bridge for the passage of locomotive trains. Rail tracks will, however, be laid over it, on which will be passed baggage and freight cars by horse or stationary steam power. Its length will be only about 800 feet, and it is to be presumed that railway passengers will much prefer crossing it on foot to any other mode because of the more satisfactory opportunity thereby afforded of contemplating the sublimity of the structure, and the magnificent gorge and torrent spanned by it.

This beautiful line drawing of Roebling's great suspension bridge at Niagara Falls appeared in a French publication. This drawing illustrates a cross-section of the structure showing the rails on the top deck and the highway underneath. Note all the extra bracing cables tied to the bank, thus giving the bridge the required stiffness for train travel.
— SMITHSONIAN INSTITUTION

Huge pin-connected braces anchor the supporting cables into rock to a depth of 25 feet on both sides of Niagara Gorge. (BELOW) A single track runs through the upper deck of the bridge, supported by great towers of stone and miles of wire rope cable. — BOTH SMITHSONIAN INSTITUTION

Obviously, the reporter of that account had not bothered to consult with John Roebling or he would have been set straight. But it is interesting to note how widely held was the idea that it was impossible to build a suspension bridge over which trains — and locomotives — could travel.

Roebling kept in constant touch with his plant near Trenton, giving them the details of the machinery installations that he would need, ordering the necessary wire rope and firing off scores of detailed instructions.

By their September 11, 1852 issue, *Scientific American* had, perhaps, a clearer idea of what the bridge would be:

> ....The anchorage will be formed by sinking eight shafts in the rock, 25 feet deep, at the bottom of which will be massive cast-iron plates, firmly held down by solid mason work. Saddles of cast-iron will support the cables on the towers, capable of supporting the pressure of 6,000 tons. The towers are to be 60 feet high, 16 feet wide at the base and 8 at the top. Weight of timber in the bridge, 910,130 lbs.; wrought iron and suspenders, 115, 120 lbs.; castings 44,333 lbs.; rails, 66,770 lbs.; cables between the towers, 335,400 lbs. When the whole is covered with a locomotive and train of cars, it is estimated that it will have to sustain a weight of 1,273 tons, which amount of burden, though not likely often to occur, is less than is provided for. It will be the longest railroad bridge between points of support in the world.

When Roebling heard of the collapse of Ellet's bridge at Wheeling, he immediately rushed an order to his Trenton works for additional wire rope which would supplement his system of bracing on the Niagara span. He also added inclined stays below the floor of the bridge in order to resist wind uplift, the force that had taken the Wheeling suspension bridge.

In June 1854, the cables and suspenders had been erected and Roebling recorded the start of work on hanging the roadway beams, high above the whirling rapids: "We commenced suspending beams last Saturday and will finish today — all done in three days. I have the satisfaction of seeing my suspenders come out perfectly correct, so much so that it will be hardly necessary to touch a screw."

Two days later, elated at how well all the parts of the bridge were going together, he wrote: "My bridge is the admiration of everybody; the directors are delighted. The woodwork goes together in the best manner. The suspenders require scarcely any adjustment at all."

As the country weathered a financial storm at the end of 1854, the bridge, too, proved that it could withstand the forces of nature, Roebling wrote from Suspension Bridge, New York: "We had a tremendous gale for the last 12 hours; my bridge didn't move a muscle."

Hanging the wooden trusswork for the railway and the roadway decks continued into the winter of 1854-55. The rails were spiked tight to a raised platform on the upper bridge floor. This deck was trued up with a slight camber or "hump" so that the sag would be counteracted when a train crossed.

In March 1855, the Niagara bridge was completed; the first successful railway suspension bridge in the world. On March 16, 1855 the first train passed over the completed structure — and it had the distinction of being the first train in history to cross a bridge suspended from wire cables. It was as a result of the Niagara bridge that the new concept of stiffening trusses received its first full expression. Filling the vertical height between the upper and lower decks of the span, this trussed bracing was 18 feet deep. The amazing success of this innovation established it thereafter as a recognized essential feature in the design of suspension bridges.

This summertime view looks out on the Roebling suspension bridge and Niagara Falls, Ontario, on the Canadian side of the bridge. From this angle, the viewer can get a good idea of just how magnificent this bridge was, and of its size. — SMITHSONIAN INSTITUTION

This side view of the bridge itself, gives a good idea as to all the bracing required to keep the bridge rigid. The railroad was on the top deck, and all pedestrians and wagons were restricted to the lower deck. — SMITHSONIAN INSTITUTION

Roebling described the opening day in a letter as follows: "Last Sunday I opened the bridge for the regular passage of trains. The first one was the heaviest freight train that will ever pass, and was made up on purpose to test the bridge. With an engine of 28 tons we pushed over from Canada to New York 20 double-loaded freight cars, making a gross weight of 368 tons; this train very nearly covered the whole length of floor between the towers.... no vibrations whatever. Less noise and movement than in a common truss bridge... Yesterday the first passenger train from the East with three cars, crowded inside and on top, went over in fine style. Altogether we passed about 20 trains across within the

By train over the Niagara Gorge! The Roebling suspension bridge quickly became a tourist attraction and photographs or illustrations of the famous structure were used in many a travel brochure and advertising broadside. Now a passenger could travel between New York and Chicago over two routes and with the option of stopping off at Niagara Falls. — SMITHSONIAN INSTITUTION

last 24 hours. No one is afraid to cross. The passage of trains is a great sight, worth seeing it."

The span had cost $400,000, over twice the original estimate, but the owners were pleased with the results.

The chief supporting members of the 821-foot bridge were four 10-inch cables, two to a side, one for the upper deck, the other for the lower. Each cable contained 2,640 wires, laid parallel and tightly bound into a functioning unit. In addition to the vertical suspenders, there were radiating stays to provide aerodynamic stability and to increase the rigidity of the whole structure.

As it should, the bridge drew international attention. It was featured on various pieces of advertising such as: railroad timetables and posters (the "Niagara Gorge Route"), boasting through trains from the East, New York and Boston, to Detroit and Chicago, as well as to Toronto. Advertisements made a point of stating that "all trains stop at Suspension Bridge." A town grew up on the American side and became Suspension Bridge, New York, now a part of Buffalo. But to railroaders, it is still Suspension Bridge and the route still figures importantly in through freight movements.

Shortly after the bridge was opened to rail traffic, the track across the structure was gauntleted in order to cut down on switching at both ends. A double track on each end formed a single track across the bridge. Note the stub switch in the foreground on the American side of the bridge. — SMITHSONIAN INSTITUTION

The book *The American Railway,* published in 1888, included these two line drawings of the towers of Roebling's suspension bridge. (LEFT) The original stone tower. At the right, the new iron towers. The iron towers were built in front of the stone towers, hence traffic over the bridge was not interrupted during the work. The suspension bridge remained until 1897 when a larger, heavier steel structure replaced it.

With its shapely stone towers rising from the high cliffs of the Niagara Gorge, and with the long graceful sweep of its cables, the span piqued the imagination and pride of Americans and Canadians alike and served as one more idealogical as well as physical link between those two great countries. In 1860 the Prince of Wales, on his American tour, made a special point of visiting Roebling's bridge. He had his train stopped on the span while he walked over the famous structure examining its construction. It had become the most talked about bridge in the world.

The bridge carried progressively increasing weights of trains and locomotives for more than four decades. After 26 years of service, the wooden suspended structure was replaced by iron and steel trusses. After 30 years, it was found that the saddle rollers at the tops of the towers lacked the necessary freedom of motion and that strains had affected the stonework. So iron towers were erected to support the bridge — all without any interruption of crossing traffic.

According to *Leslie's Weekly:*

> The official testing of the Niagara Suspension Bridge began April 17, 1877 and lasted four days.
> The original test of the bridge in 1853 was done with a pressure of 326 tons.
> In the 1877 test the pressure was 450 tons — twenty loaded freight cars and engine — and the deflection was the same proportion as in the first test. The stability of the bridge was completely demonstrated. Regular traffic was resumed on April 25th.

Occasionally, in the succeeding years, chord members of the new metallic stiffening truss would break due to the increasing locomotive weights, but repairs were made without any interruption of service.

Finally in 1897, when weights and sizes of locomotives had increased far beyond the imagination of bridge builders of the 1850's, the decision was made to replace the suspension bridge. A double-decked steel arch bridge, anchored into the rock walls of the cliffs, grew over and around the old suspension bridge. During the erection, the old bridge was kept in service for as long as possible.

The arch bridge, which carries on the proud tradition today, is 780 feet long (See chapter on Steel Arch Bridges) and is built very close to the original site. Some of the original stone foundations of the suspension bridge may still be seen and, also, one may still view the boiling waters of the mighty Niagara and the whirlpool below.

John Roebling, of course, went on to greater heights when he constructed a suspension bridge in Kentucky over the Kentucky River, and then to his great work: the Brooklyn Bridge, which was completed by his son, Washington Roebling, after the tragic death of his father.

The west bay crossing of the San Francisco-Oakland Bay Bridge is one of the world's great suspension bridges. It is also worthy of note here because it originally carried interurban railway traffic on its lower deck. Trains of the Key System, the Sacramento Northern and Southern Pacific's electric operation, Interurban Electric, all carried thousands of commuters in their day. In the view above, a Key System train rolls toward Treasure Island station. — WILL C. WHITTAKER

It is interesting to note that the Brooklyn Bridge, also accommodated tracks for streetcars. When the San Francisco Bay Bridge was built, it too, had rails, for the Key System electric commuter cars, the suburban electric trains of the Southern Pacific and the Sacramento Northern interurbans. Now neither of these two suspension bridges

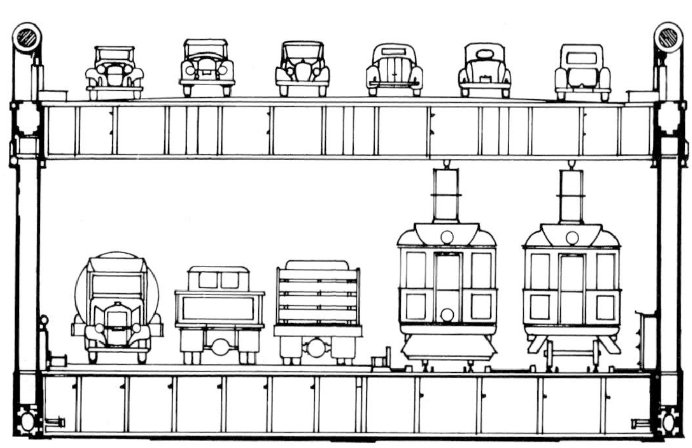

The San Francisco-Oakland Bay Bridge looking from the San Francisco side toward Treasure Island — RICHARD J. COOK (RIGHT) This drawing shows how the railway fits into the lower deck of the bridge. The Bay Bridge, 43,500 feet long, was built between 1933 and 1936 with November 12, 1936 being the official opening day.

Philadelphia's 1,750 foot Benjamin Franklin Bridge, completed in 1926, was the longest suspension bridge in the world when built. The lower level was constructed to accommodate rapid transit cars along the outside of each side of the bridge. One car may be seen passing the east tower. — RICHARD J. COOK

carries rail traffic, but there is one that does: the great Benjamin Franklin Bridge over the Delaware River linking Philadelphia with Camden. On the single deck trains of the Delaware River Port Authority's PATCO (Lindenwold Line) carry commuters in increasing numbers to South Jersey communities.

A PATCO rapid transit train greets the sun over the Delaware River, as it heads east across the only suspension bridge in North America which presently carries rail traffic. — RICHARD J. COOK

The first important American bridge of steel was the famous Eads Bridge at St. Louis. In this scene, a passenger train is visible near the center of the cantilever arch. — RICHARD J. COOK

4
The Eads Bridge

As the nation moved westward, the railroads extended out from Baltimore, Philadelphia and New York, and crossed rivers on stone arches and wooden or iron trusses. The Ohio River barrier gave way to such bridges — and it was on to St. Louis and the greatest river obstacle of them all: the Mississippi.

In time, even this barrier was overcome. It was James Buchanan Eads who bridged the mightiest river at St. Louis and literally opened the West to rail traffic from the East. Until the great Eads Bridge was opened in 1874, all trains terminated at East St. Louis. All passengers, freight, mail and express crossed the Mississippi by means of Wiggins Ferry.

Eads, a midwesterner and an authentic American genius, was slim, leathery, highly opinionated, and disliked by many, but nevertheless, had ability. He was born in Lawrenceburg, Indiana, in 1820, and went, when still a boy, to live in St. Louis. There he learned to know the Mississippi, to respect it and to love it. At the age of 18 he became a purser on a riverboat and for the next four years, steadily traveled the river between St. Louis and New Orleans.

In 1842 he invented a diving bell with which he salvaged sunken steamers and, as a result he made a fortune. In 1845 he established a factory at St. Louis for the manufacture of glass. With his accumulating wealth, he was able to build and pay for a fleet of ironclad wooden gunboats. During the Civil War these played a decisive role in defeating the Confederates on the Mississippi. He had, by that time, a well-earned reputation as a entrepeneur and had an extensive knowledge of the great river.

Because of his work up and down the river with the diving bells, Eads was said to know more about the Mississippi's treacherous currents and the character of its bottom than any man alive. He was a man of great courage, forcefulness and unbridled self-confidence., Even though he had had no formal training as an engineer, nor had he ever built a bridge before, he wanted to build a bridge over the river he knew so well. Charles

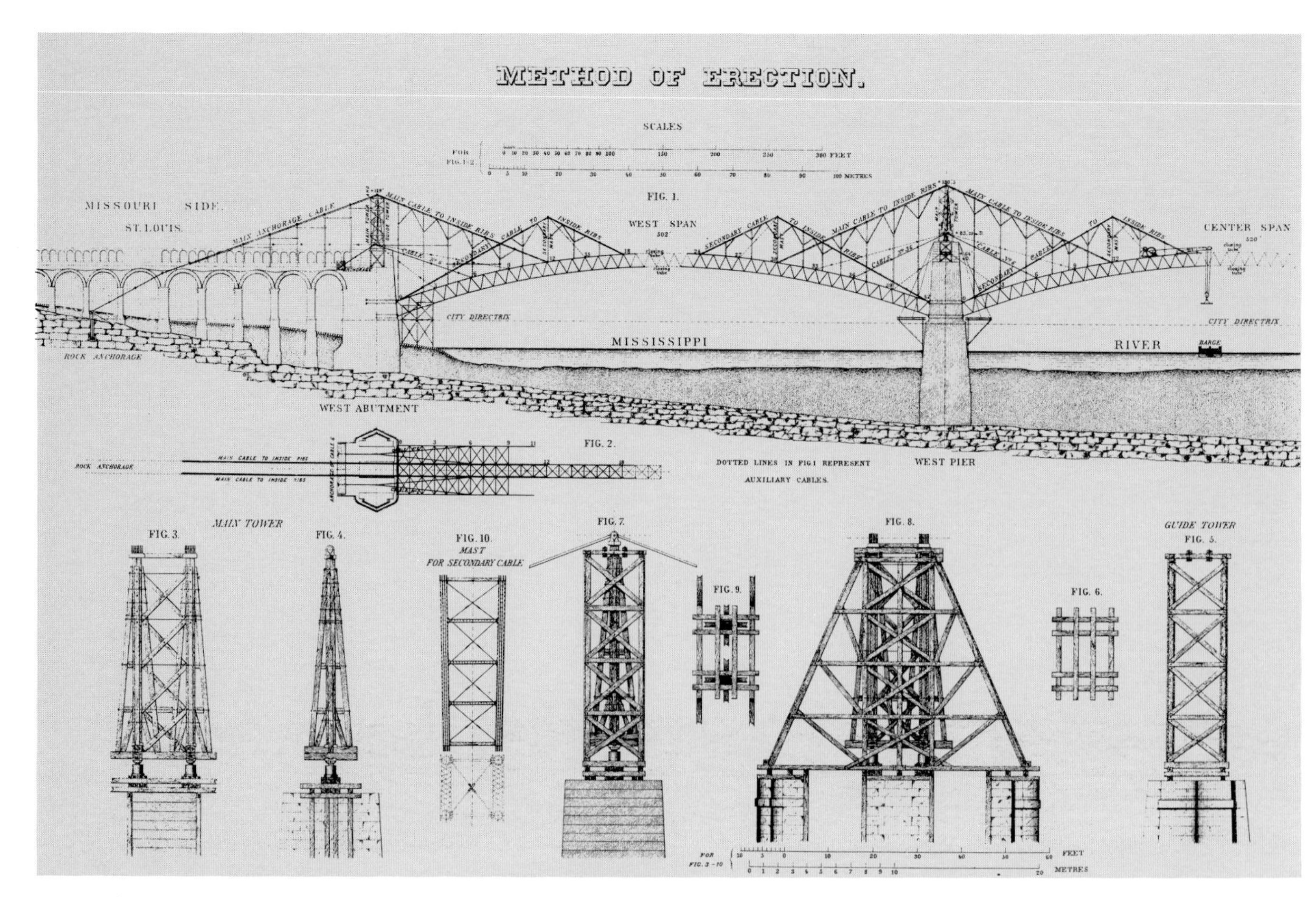

As a work of structural art, the Eads Bridge remains a classic. Its arches of special crucible steel were erected without falsework in the river. It was the first major bridge in the United States using the cantilever method of extending the arches out from either side and meeting in the middle. — SMITHSONIAN INSTITUTION

Ellet and John A. Roebling, in the 1850's, had both prepared plans for a suspension bridge at St. Louis which were turned down.

The city fathers chose Eads to build the bridge because of his intimate knowledge of the river. So, in 1867, it was up to him to find a way to rest bridge piers far below the irregular and turbulent current of a river loaded with sediment and choked with heavy debris in the rainy seasons, not to mention the hazards of drifting ice in the wintertime. A formidable task indeed. This great river could be a destructive force of immeasurable power. Although river pilots could read the surface with remarkable skill, only Eads knew firsthand the changing, shifting bed. He had seen the riverbed change from 20 to 100 feet where obstacles influenced the underwater currents.

The great need was for a bridge to carry both railroad and highway traffic over the river without interfering with steamboats. Eads' idea for spanning the river was to have a mammoth bridge, built with arches of steel set on stone piers. It would have just three of these tubular steel arches, the biggest of which, the center span, would be longer by several hundred feet than any arch of the time.

To found the piers on bedrock, he built a cylindrical iron-shod caisson of massive timbers heavily reinforced with iron bands, 75 feet in diameter, with a working chamber of eight feet in depth. Within this huge enclosure the masonry pier was built up, the weight of the masonry forcing the cutting edge into the riverbed. As the caisson sank lower into the water, successive rings were added to it. Sand pumps drew out the excavated material. Meanwhile, men and supplies moved in and out of the working chambers through a cylindrical opening left in the center of the masonry.

Five months of pumping and excavation

With the most difficult work of sinking and erecting the piers completed, progress went forward on the upper structure of the Eads Bridge. The scenes above, and to the left, show the erection of the bridge by the cantilever method. The arch structure consists primarily of four vertical pairs of tubes, with the pairs set side by side across the width of the deck. (BELOW) The architect's rendering of the Eads Bridge dramatized the height of the structure and the fact that steamboats were not required to lower their stacks in order to pass through an arch. This drawing includes statues atop each pier, an adornment never installed. — ALL SMITHSONIAN INSTITUTION

An 1871 architectural drawing showing the west abutment. Note the train located on the lower deck. (BELOW) A photograph of the newly completed Eads Bridge, the first spanning the Mississippi River, shows the blending of the bridge with river commerce. The new bridge provided a stimulation to both rail and water commerce. Within a decade, St. Louis would become the second largest rail terminal in the United States. — BOTH SMITHSONIAN INSTITUTION

finally resulted in the long-awaited goal: the foundation rock. But along the way a serious problem had arisen. The increased air pressure on the construction workers resulted in cases of painful "bends" which began to appear at the 60 foot level below the surface. At 76 feet, the first case of paralysis and crippling appeared and, when one of the foremen got sick, Eads decided to shorten the shifts inside the caisson. The men would only stay down for four hours and rest for eight hours. Then, at greater depths, the shift on duty was reduced to two hours. Still, as the excavation went deeper, the mysterious trouble increased.

Up to March 19, 1870, there had been only a few mild attacks of the mysterious "caisson disease," but now things went very wrong. One man came up from the air chamber declaring that he was not feeling well and, 15 minutes later, died. The same day another died at the hospital. With more deaths occurring, Eads frantically ordered that, as the men came up, the air pressure should be reduced more slowly. Thus with the shortened working shifts plus the new technique of air pressure reduction and slower emerging time, the illness came under control. The work continued. By 1873 the piers and masonry arch approaches were completed.

The steel tubes that constitute the arch ribs were erected by the method known as cantilevering under backstays: a completed arch segment was held in place against the pier by cables that extended from the free end to towers on the next pier or shore abutment. The stays held the lengthening segments of the two half-arches as cantilevers until the arch was closed at the crown. The arches were built simultaneously from the pier faces.

The arch structure of the Eads Bridge is unique. It consists primarily of four vertical pairs of tubes, with the pairs set side by side across the width of the deck. All arches are fixed at their ends. There are three spans; 520 feet for the center and 502 feet for the flanking arches.

As a work of structural art, the Eads Bridge remains a classic. Besides that, it established several engineering "firsts" in bridge building. The first major bridge to be erected by the cantilever method, it was also the first bridge

A recent photograph of the east abutment of the Eads Bridge shows that time has not altered or taken its toll on the solid and venerable structure which sees daily heavy rail and automotive traffic. — RICHARD J. COOK

Plate XIII

SECTION OF EAST PIER AND CAISSON

ON LINE AB, PLATE VII.

SHOWING THE INTERIOR OF THE MAIN ENTRANCE SHAFT AND AIR CHAMBER AND THE WORKING OF ONE OF THE SAND PUMPS

For the deep foundations of its piers, the pneumatic caisson process was used for the first time in America. This view shows a section of the east pier and caisson and the interior of the main entrance shaft and air chamber, plus the working of one of the sand pumps. — RICHARD J. COOK COLLECTION

in the world to employ chromium alloy steel for toughness and corrosion resistance. The chromium steel hingeless arch ribs have 100,000 pounds ultimate strength and 40,000 pounds per square inch elastic limit. Fifty years later, tests were made on the condition of the steel arches and they were found to be without flaws.

No previous arch or even truss spans 500 feet long were in existence at the time. Nowhere had bridge piers been built from so deep a river foundation. And the pneumatic caisson, invented in France in 1868, was first put to a real operating test in the construction of the Eads Bridge.

The Eads Bridge was completed in 1874 at a cost of $6,536,729.99. On July 2, 1874, there was a public test of the bridge which was witnessed by thousands of citizens. Fourteen heavy locomotives crowded with people traveled over the bridge. As the engines slowly puffed across, a great shout arose from the people. Their bridge was a success.

From that time on rail transportation flowed smoothly into St. Louis from the East, no longer hindered by the "Father of Waters."* The bridge was, indeed, most important in opening up the West as St. Louis soon became the second greatest rail terminal in the United States.

Its four-lane upper deck carries a heavy traffic of trucks, buses and automobiles. On the second deck are two tracks used by the Terminal Railroad of St. Louis and by Amtrak passenger trains.

James Buchanan Eads is the only engineer in the Hall of Fame and his bridge, also, has been honored. It has been proclaimed a national historic civil-engineering landmark by the American Society of Civil Engineers.

*The Mississippi River was known as the "Father of Waters."

Pennsylvania Railroad GG-1 No. 4852 with the northbound *Legislature*, crosses the Susquehanna River bridge, a deck truss, at Perryville, Maryland, June 1968. This route now forms the Amtrak Boston-Washington corridor. — H. H. HARWOOD, JR.

5

The Truss Geometry at Work

The truss is probably the most commonly used construction form for a railroad bridge. It acts to equalize the load in the most economical manner and has appeared in many forms. Basically, it uses a pair of ladder-like frames, running from abutment to abutment (or from pier to pier) with the roadway beams laid crosswise between them, either on the top or bottom horizontal member. Adding diagonal members to the framed structure makes it a truss and greatly strengthens it. Thus, a rectangle bisected by a diagonal becomes two triangles, and the triangle is the geometric figure that is rigid. (Though a triangle built of three rigid pieces pinned together is theoretically indestructible, in reality, the strength of materials and joints must be considered. Pressure applied to any point of the triangular structure will produce tension in one or more members and compression in the others, and this is the basic concern of bridge engineering.)

If the roadway is laid on the top horizontal member — which is called a chord — the bridge is a *deck truss*. If laid on the bottom member, it is a *through truss*.

A New England-born carpenter, Theodore Burr, a relative of the famous Aaron Burr, was the one who brought bridge design closer to the needs of the railroad. Burr, who spent most of his life in Harrisburg, Pennsylvania developed a basic flat truss; its top and bottom chords were straight and parallel, and either could support a level railroad track. The design inspired confidence in other bridge builders and was widely used throughout the 19th century.

Other early truss designs used on railroads were the Bollman, Town lattice, Pratt, Fink, and Howe. Ithiel Town, a New Haven architect, patented his truss (1820) which served the needs of many railroads in their earliest years. Town filled the space between the top and bottom chords of the truss with a latticework of diagonal pieces. The result was not only sturdy, but also easy and cheap to produce. He charged a dollar a foot for every licensed bridge built under his patent — and two dollars a foot whenever he found one built

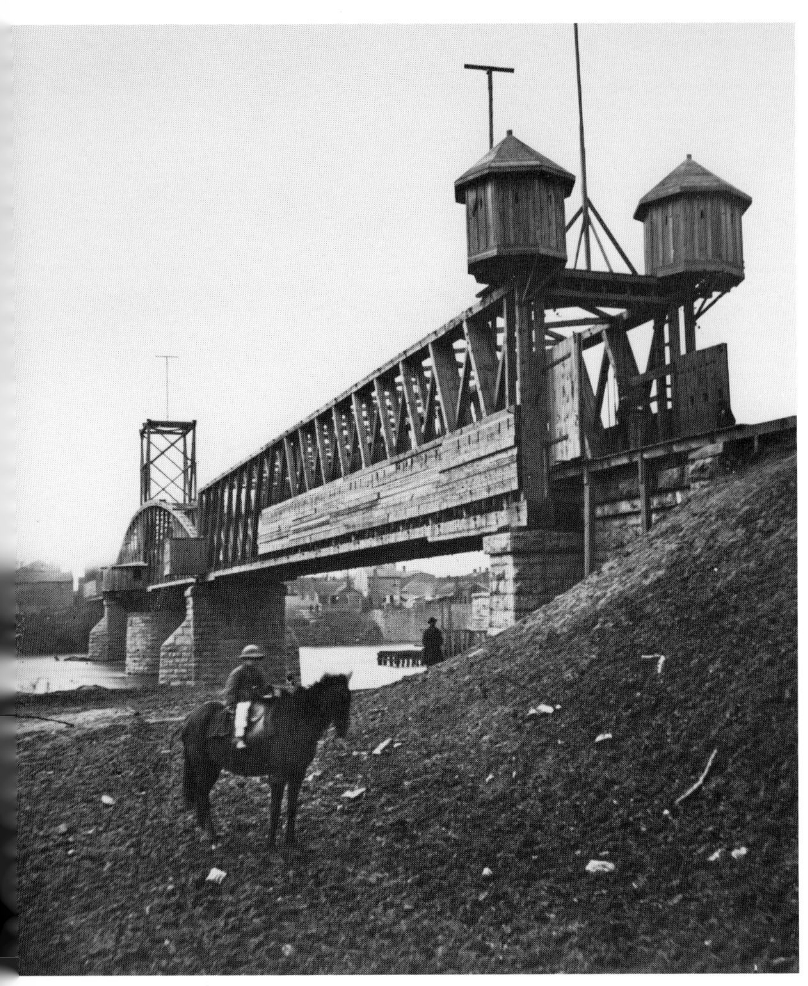

For over a century, wood had been the principal material used in the construction of railroad bridges. Wood resisted both compression and tension and a single piece was suitable for use as a beam. The greatest weakness of timber as a bridge building material lay in the difficulty in connecting together separate pieces. Another material, such as iron or steel, was used to fasten timbers together to form trusses: but even then it was impossible to make a joint that would be as strong in tension as timber. However, wood was more suitable for compression than tension, and some of the old timber trusses had iron tension members. Early wood bridges were built with untreated wood and, unless covered, had a short lifespan. (ABOVE) Construction workers appear to be in the final stages of construction of this timber-arch bridge at Rockford, Illinois, in 1869. — SMITHSONIAN INSTITUTION (LEFT) This Civil War era arch bridge also sports sentry boxes used for the protection against Confederate bridge wreckers. The location is unknown. — LIBRARY OF CONGRESS

Brigadier General Herman Haupt, chief of construction and transportation of the U.S. Military Railroad during the Civil War, performed miracles in the construction of prefabricated railroad bridges. In the above scene, Union officers and their families pose for a portrait in 1862 in front of a Haupt arch and truss bridge over the Rappahannock River. — LIBRARY OF CONGRESS

In building the Southern Pacific from northern California to Texas, it was necessary to bridge the Colorado River at Yuma. During June 1877 the SP built this 667-foot "straining beam truss" bridge consisting of six 80-foot spans and a swing span of 93½ feet on the Arizona side and out of the picture. This allowed flat bottom river steamers to follow the deepest channel of the Colorado River during periods of high water. — DONALD DUKE COLLECTION

Few people, if any, outside of the engineering profession, appreciate the debt we owe to a highly efficient wood bridge type which was used most extensively in early railroad building in the United States. That is the Howe truss bridge. Because of the availability of timber in the Pacific Northwest, this type of bridge was used extensively there. The four views here show a perfect example of a Howe truss over the Noosack River at Everson, Washington. The bridge, built in 1933, suffered damage by fire a few years ago, and now is listed as a landmark by the Washington State Historical Society. It was originally built by the Milwaukee Road, but is now owned by the Burlington Northern. — ALL RICHARD J. COOK

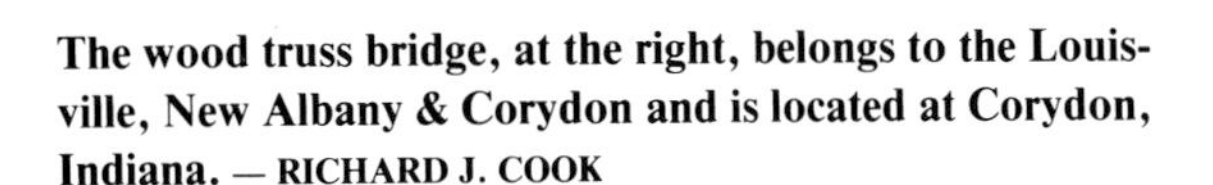

The wood truss bridge, at the right, belongs to the Louisville, New Albany & Corydon and is located at Corydon, Indiana. — RICHARD J. COOK

The Covered Truss Bridge

Railroad and highway bridges were covered in order to protect the untreated wood from the elements. But often the value of this protection was questionable due to the hazard of fire caused when the hot cinders from the stacks of the steam locomotives would hit the roof and destroy the bridges. Treating of bridge timbers with creosote eliminated the need for covering bridges. A few classic covered rail bridges survived through the 1950's. (LEFT) The covered bridge on the narrow-gauge East Tennessee & Western North Carolina was famous. — JOHN KRAUSE (BELOW) Built in 1937, the Milwaukee Road covered bridge over the Stuck River at Sumner, Washington, looked like something from out of the past when photographed in 1968. — H. E. BENDER

The Milwaukee Road covered bridge over the Snoqualmie River at North Bend, Washington, was built in 1910. When H. E. Bender photographed it in 1968 the boards on the side were falling off.

Vermont has been famous for its many covered bridges. In the above view a St. Johnsbury & Lamoille County Railroad diesel rolls a freight through a beautifully covered bridge. — RICHARD J. COOK COLLECTION

These three views are of a Bollman truss bridge at Savage, Maryland, built in 1852 and moved to its present site in 1889. The structure, no longer connected by railroad, is owned by Howard County which plans to restore it to its original appearance. — ALL H. H. HARWOOD, JR.

without his permission. The bridge design was popular, but the increasing weight of locomotives eventually proved it inadequate. When a Town truss over Catskill Creek in New York State collapsed in 1840, the popularity of the design quickly dwindled.

It was during the 1840's that cast iron elements began to be included in wooden truss bridges. Because of its brittle nature, however, cast iron construction was often aided by wrought iron members. At first diagonal tie rods were used for greater tensile strength. Although wrought iron was more expensive, it was also more elastic. Eventually wrought iron was used for entire structures; cast iron for railroad bridges soon fell into disuse.

Wendell Bollman, who became a foreman of bridges for the Baltimore & Ohio Railroad at an early age, replaced a portion of the Wernwag wooden bridge at Harper's Ferry, West Virginia, with an iron truss bridge of his own design. In an official test, three locomotives rolled across the 124-foot span, distributing a weight of more than a ton per foot of bridge, yet causing a deflection at the bridge's center of less than an inch and a half.

A giant leap forward in the art of bridge building — and just in time for the increasing demands of heavier, longer trains — can be traced back to 1855 when the English scientist, Henry Bessemer, invented a practical way to make steel. His Bessemer process brought the possibility of making bridges of steel into the realm of reality. Soon there were new designs and greater and longer bridges being found all over the world.

Huskier bridges meant that longer trains pulled by heavier locomotives could move goods with greater economy. The freedom of bridge engineers to design a great bridge of steel that would withstand thousands of tons in movement, coincided with the needs of the industrial revolution. A transportation system evolved which thrust ribbons of rail to all corners of the continent. American, Canadian and Mexican rail systems were in their prime. Great rivers and canyons, formerly unbridgeable, were conquered, making the shortest routes practicable.

New phrases, such as "new route," "low grade line," "cut-off," or "airline" came into the railroaders' language. New methods of grading, cut-making and tunnel-boring were

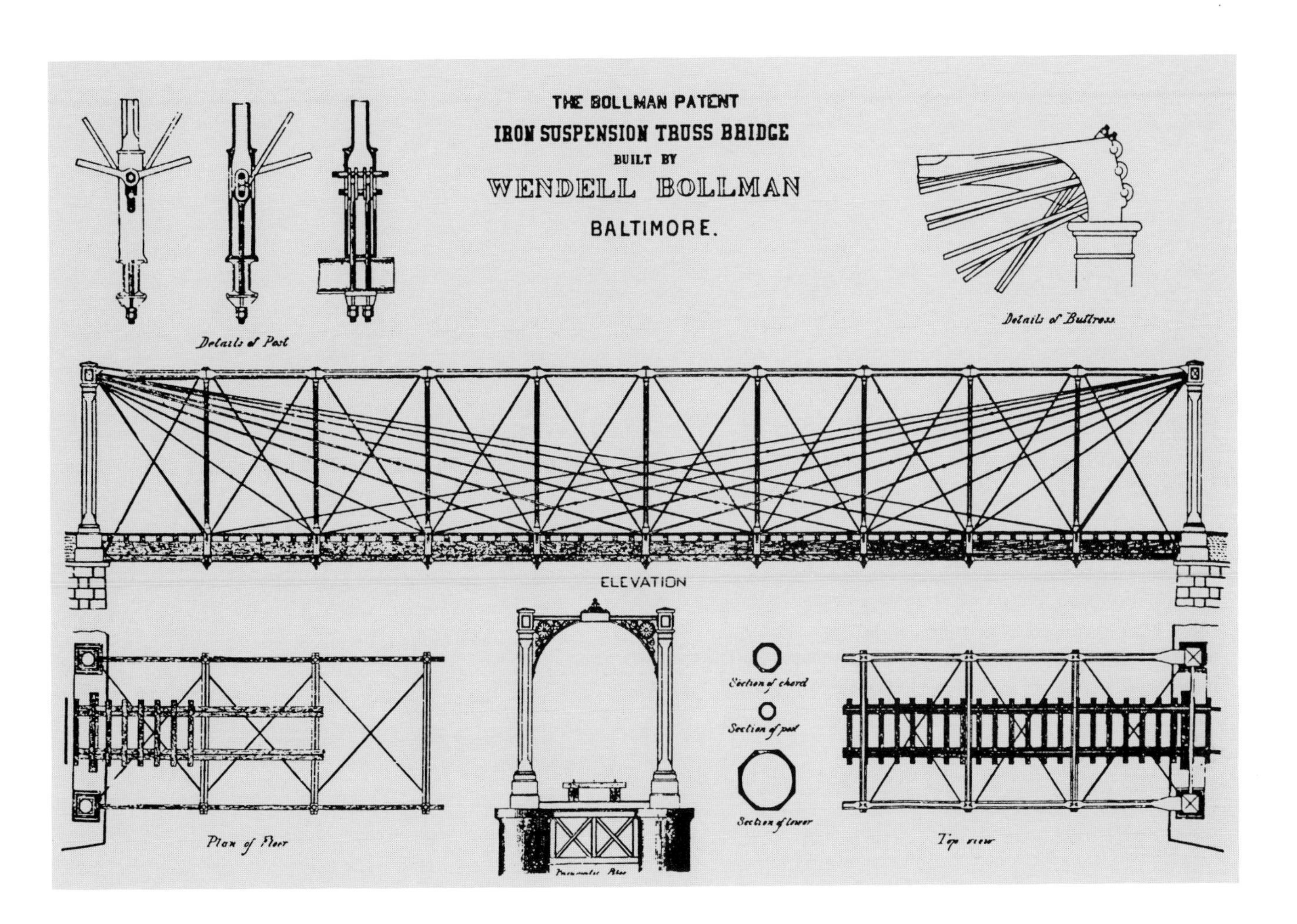

Wendell Bollman, a onetime carpenter, became the foreman of bridges on the Baltimore & Ohio Railroad. As many of the wood bridges came up for replacement, Bollman designed an iron truss of his own patent which is shown in the above view. This unique spiderweb design became a standard on the B&O railroad.

combined with the bridge-making art. The great railroad age had arrived.

Some notable railroad truss bridges down through the years have been the Fink all-iron bridge for the Baltimore & Ohio over the Monongahela River at Fairmont, Virginia (later West Virginia), 1852; the Louisville & Nashville bridge over the Green River in Kentucky, 1859 (A Fink bridge, it had five spans totalling 1,000 feet and became the longest and most spectacular iron bridge in America at that date); J. H. Linville's long-span truss over the Ohio River at Steubenville, Ohio, for the Pittsburgh, Cincinnati, Chicago & St. Louis (later Pennsylvania Railroad), 1864; the great Quebec 1,800-foot cantilever near Quebec City, 1917; and the Huey P. Long Bridge over the Mississippi River at New Orleans, 1936.

Linville, who successfully challenged the Ohio River at Steubenville, later became president of the Keystone Bridge Co. He went on to other ambitious projects, including two more bridges over the Ohio River: the Baltimore & Ohio at Parkersburg, West Virginia, and the Cincinnati Southern at Cincinnati. The Cincinnati bridge, built in collaboration with L.F.G. Bouscaren, was completed in 1877 to connect the Kentucky coal fields with the Great Lakes industrial region. His truss span of 517 feet was the longest span in the world at that time.

Writer Carl Condit described the importance of Bouscaren's engineering plans and specifications as:

> . . . the prototype of all bridge specifications . . . the first to embody in detail all the criteria of design, loading, material, workmanship, and safety that the builder follows in the erection

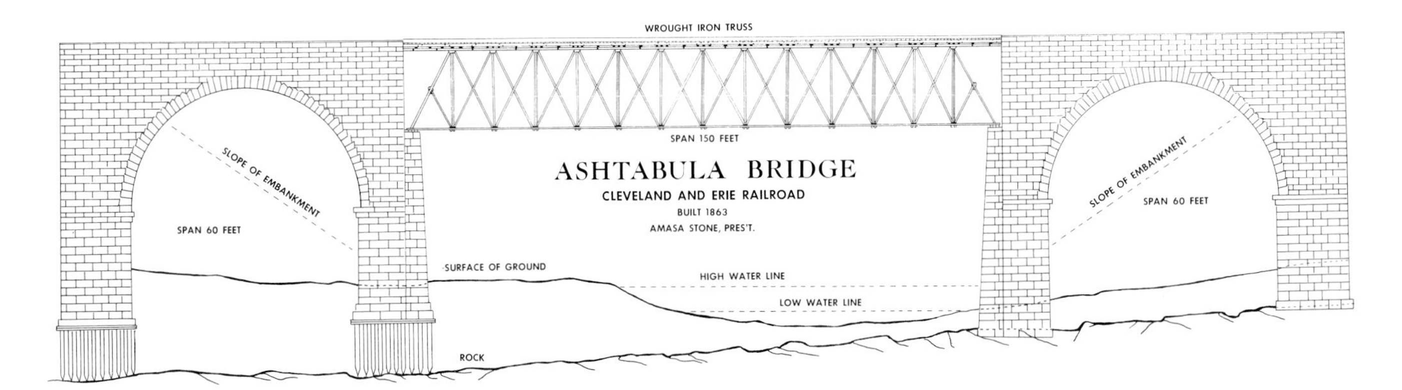

An artist's drawing of the ill-fated Ashtabula bridge. The collapse of this Howe truss in 1876 resulted in one of the worst railroad bridge disasters of all time. The tragedy resulted in two suicides, besides the death toll of about 150 passengers aboard the train. — BOB LORENZ

of a major railroad span... The whole modern program of construction was prefigured in the Cincinnati project—geological and hydrographic survey of the site, determination of loads, stress analysis and calculation of the size of truss members, preparation of working drawings and specifications, competitive bidding, the testing of sample materials, inspections at the mill and at the site, and the training of mill hands and construction workers. With this program as a guide, the builder could work with confidence and the public would expect an end to the grim series of disasters that once marked the history of the iron bridge.

One of the worst railway bridge disasters of all time was the collapse of the Howe truss, a 165-foot wrought iron span over the Ashtabula River at Ashtabula, Ohio, on the Lake Shore & Michigan Southern.

On the night of December 29, 1876, an eleven-car passenger train pulled by two locomotives began to cross the bridge at 15 miles per hour. It had been snowing all day and with the snow still pelting down, No. 5, two and one-half hours late, moved slowly forward. As the engineer of the first locomotive reached the center of the span, he felt the bridge begin to sink. He opened the throttle to its widest notch and managed to get his own engine across, but the other engine and the rest of the train, together with the entire span, went crashing into the ravine, and into the icy waters of the river 70 feet below. The train promptly caught fire and those who were not victims of the tumultuous crash itself perished in the flames of the crushed cars.

For many years the Office of Safety of the Federal Railroad Administration listed the death toll as 84. Later it was raised to 92, based on the death claims filed with the railroad.

The conductor, Bernhard Henn, testified that there were 131 persons aboard the train, supporting the railroad's claim that the death toll was only about 60. The passengers, though, had said that Henn had told them shortly before the accident that there were more than 160 aboard. Many passengers felt that the total was more than 200, indicating a death toll of up to 150.

The Ashtabula tragedy was highly publicized and aroused a storm of protest. Blame was first laid on the bridge itself with the claim that it had been cheaply constructed. But that was disproved. Then the blame was shifted to the man responsible for inspecting the bridge, Charles Collins, the railroad's chief engineer. He was so distressed by the charges that he committed suicide.

The coroner's verdict put the blame on Amasa Stone, president of the Lake Shore Route. Stone was one of five brothers-in-law of William Howe, inventor of the span. Stone was blamed for building a bridge "without the approval of any competent engineer and against the protest of the man who made the drawings... assuming the sole and entire responsibility himself." Previously, Collins had objected strenuously to the design of the bridge and refused to supervise its construction. According to the testimony, he had been overruled by Stone.

Stone insisted on building an iron bridge

using his brother-in-law's Howe truss design. Besides, it was pointed out, to have built a bridge of stone would have cost an additional $18,000. Both Stone and his design were eventually condemned by the American Society of Civil Engineers. In 1883, troubled by a series of business reverses, driven to distraction by insomnia, Amasa Stone put a bullet through his heart.

Actually, however, no one could establish exactly why the Howe span had fallen. The soundest theory was that a car or two might have become derailed just before reaching the bridge. The derailed wheels then would have struck the flooring causing the beams to give way, pulling both lines of trussing inward at the top. This could have happened to any bridge of the period with equally disastrous results.

This ancient wood truss bridge served the Canadian Pacific Railway for many years. A local train on the Chipman-Norton branch crosses this classic structure located north of Cody, New Brunswick. The bridge was still in service when photographed by H. H. Harwood, Jr. in 1959.

This old iron truss bridge, located at Stewartstown, Pennsylvania, was erected by the Northern Central Railroad, and later formed a part of the Pennsylvania Railroad. — H. H. HARWOOD, JR.

A freight train rolls across the International Bridge, spanning the Niagara River, between Fort Erie, Ontario, and Buffalo, New York. — RICHARD J. COOK

The truss is divided generally into three different types: the *simple truss,* and *continuous truss* and the *cantilever truss.*

The simple truss, which represents the majority of railroad truss bridges, reached its greatest length in the Chicago, Burlington & Quincy Bridge at Metropolis, Illinois. Here, in this little Illinois river town, which once boasted as being the "home of Superman," this big but somber and conventional bridge united a long branch line of the Burlington (now Illinois Central Gulf) with the Nashville, Chattanooga & St. Louis Railway (now Louisville & Nashville) at Paducah, Kentucky, and thus established a new freight route that joined the midwestern prairies with the southeastern states and the Gulf Coast. The overall length of the river crossing at Metropolis is nearly 3,500 feet, divided into seven spans, of which the channel span has the record length of 722 feet.

The longest continuous truss was built for the Chesapeake & Ohio at Sciotoville, Ohio, during the years 1914-17, to carry heavy coal trains north over the Ohio River to docks at Toledo, and on north, east and west to steel mills in Detroit, Cleveland and Gary. This enormous structure constitutes the ultimate expression of mass and power among American truss bridges. It is the heaviest truss of its kind.

The Quebec Bridge over the St. Lawrence River is the world's longest cantilever span. This huge steel structure was finally completed in 1917, but only after two terrible disasters in which the completed center spans fell. The 1907 collapse resulted in the deaths of 75 men; then in 1916, eleven men died when the center span of the new bridge collapsed while being hoisted into place. All three of these great bridges are being treated individually in succeeding pages.

The Metropolis bridge, said to be the longest simple truss span ever built, was the most important link in the Illinois Central's Edgewood-Fulton Cut-off. Engineers Ralph Modjeski and C. H. Cartlidge used a simple Parker truss.

Built for the Burlington, 1914-17, it was later taken over by the Illinois Central. The Kentucky section of the Edgewood-Fulton Cut-off was opened on April 7, 1927, and the Illinois section opened May 7, 1928. The bridge was built at a cost of $4 million.

The overall length of the bridge is 3,500 feet, and the longest of the seven spans has a record length of 722 feet. Pier construction at Metropolis was a relatively simple matter, because it was free of the problems that are usually encountered in foundation work on broad, navigable waterways. The concrete piers rest on a thick bed of compact gravel, near enough to the surface of the riverbed so that it could be reached by open excavation within cofferdams.

At the same time that the Metropolis bridge was being erected over the Ohio River, a truss bridge of another type was being formed over the same river system, just upstream, from Portsmouth, Ohio, at Sciotoville. This was the Chesapeake & Ohio's Sciotoville continuous span bridge, a double-tracked structure built 1915-17 by Gustav Lindenthal and with David B. Steinman as designing engineer.

Two spans of 775 feet were necessary to satisfy navigation requirements and became, at the time, the longest continuous spans in the world. This bridge is a real heavyweight, and it has to be in order to handle the long coal trains that cross it from the coal fields of

The longest continuous truss ever built, consisting of two spans of 775 feet, is this monstrous Chesapeake & Ohio bridge at Sciotoville, Ohio. It is a double-track structure built in 1917 by Gustav Lindenthal and with David B. Steinman as the designing engineer. — RICHARD J. COOK

Another record-breaker, also completed in 1917, is the Illinois Central bridge at Metropolis, Illinois, the longest simple span bridge. It has seven spans, the longest of which has a record length of 722 feet. This photograph, taken at sunset along the bank of the Ohio River, shows to good advantage the structural design of this graceful structure. — RICHARD J. COOK

Breathtaking is about the only word that comes to mind when one looks through the largest truss bridge — the Chesapeake & Ohio bridge at Sciotoville, Ohio. — RICHARD J. COOK

Kentucky and West Virginia bound for the Lake Erie docks at Toledo and the steel mills of the upper midwest. The continuous truss design affords an ideal solution to the problem of the navigation requirements, yielding maximum economy of steel, maximum rigidity under rail-traffic and the important advantage of cantilever erection, leaving the main river channel unobstructed by false-work. Compared with two simple spans, the continuous design saved 20 percent of the steel required. It established the continuous design of American bridge-building practice and held the record until an even longer highway bridge of its type was built over the Mississippi River at Dubuque, Iowa. The Sciotoville bridge, however, is still the longest riveted truss bridge in the United States.

Back in the days of steam, a Chesapeake & Ohio coal train, gives one an idea of the relative size of a huge steam locomotive in contrast to the giant bridge through which it has just passed. (BELOW) The continuous truss design of the Sciotoville bridge affords an ideal solution to the problem of navigational requirements: maximum economy of steel, rigidity under rail traffic, and an unobstructed main river channel. — BOTH RICHARD J. COOK

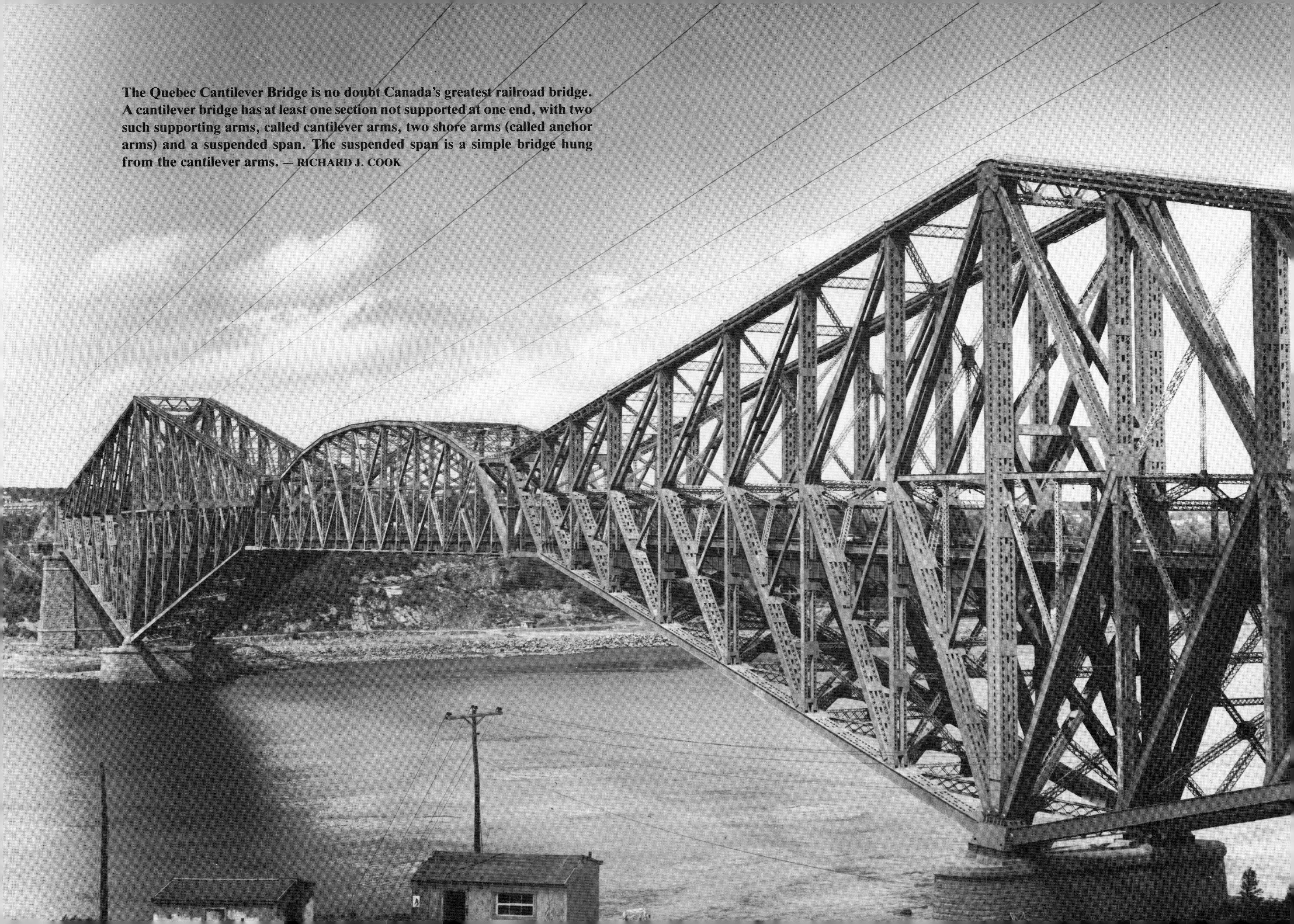

The Quebec Cantilever Bridge is no doubt Canada's greatest railroad bridge. A cantilever bridge has at least one section not supported at one end, with two such supporting arms, called cantilever arms, two shore arms (called anchor arms) and a suspended span. The suspended span is a simple bridge hung from the cantilever arms. — RICHARD J. COOK

The advantages of the continuous truss are as follows: 1) Economy of material. 2) Suitability for erection of one or more spans without falsework. 3) Rigidity under traffic. 4) Less abrupt stress-changes under traffic. 5) Elimination of expensive and troublesome hinge-details. 6) Required less extra material or hazard in erection; and 7) Safety of the completed structure.

The first great bridge in Canada, and probably still that nation's greatest bridge, encountered more hard luck than almost any structure in the western hemisphere. The Quebec Bridge, as it is called, was erected as a result of the need to cross the mighty St. Lawrence River downstream from Montreal. Canadians felt that their developing country needed a better connection to link Quebec City with the opposite shore than the ferryboats then in use. It would also speed up freight transportation to the eastern seacoast from the interior of the country.

A bridge over the St. Lawrence had been contemplated as far back as 1852, and in 1898 the Quebec Bridge & Railway Co. was formed and it made the first move toward building one. The task was formidable: crossing a wide, navigable swiftly flowing river with hard rock banks on either side. Had the proposed bridge been for highway traffic, the answer might have been to put up a suspension bridge, but the necessity here was to support heavy moving trains, thus ruling out such a structure. The structure had to be of immense strength and yet long enough to reach both shores without a center pier for support. The answer was to erect a cantilever bridge.

The term, cantilever, means a projecting arm that is either secured in some permanent fixture at the end or supported at a point between there and its outer end. When this is applied to bridge construction it denotes that a cantilever span consists of two great arms or brackets, projecting towards each other from opposite banks or piers and serving to form a bridge when united directly or by a girder. The weight of each arm is secured by what is known as the anchor span or the span between the shore and the first pier on each side, and then is securely fastened down or anchored in the masonry work of its piers.

The many years of bridge building, plus two catastrophic failures, made the Quebec Cantilever Bridge page one news for a long period of time. These photographs show the bridge in various stages of construction. — JIM SHAUGHNESSY COLLECTION

The south arm of the Quebec Cantilever Bridge was completed in mid-1907. Here something went wrong in one of the bottom chord members and the entire structure suddenly crumbled and came crashing down. — JIM SHAUGHNESSY COLLECTION

A cantilever type of bridge has many advantages, not the least of which is the possibility of its being erected without putting up falsework in the main channel. In as much as even temporary interference with navigation is now prohibited by law, this type of long span bridge has become almost mandatory.

The bridge over the Firth of Forth in Scotland held the world's record for the greatest length of cantilever span up until the Quebec Bridge was completed in 1917. It has a span length of 1,800 feet. The design of the structure, as finally built, was the work of Phelps Johnson and G.H. Duggan of the St. Lawrence Bridge Co., acting with an advisory board of five engineers, headed by Ralph Modjeski.

The final span was completed but not until great difficulties had been surmounted. In 1900, Theodore Cooper, then the most eminent builder of railroad bridges in America, was engaged to design the new record-breaking structure which would have a span that surpassed the Forth Bridge's by 100 feet, and at an estimated cost that was enormous. The designing engineers were under great pressure to utilize their utmost ability and resourcefulness to keep down the amount of steel used. Despite the unprecedented dimensions and proportions, facilities and funds for experimental investigations to guide the design were not made available.

On August 29, 1907, as the south arm of the great center span was approaching mid-channel, something went wrong in one of the bottom chord members and the entire structure suddenly crumpled and came crashing down. Up to 82 men (figures vary) were carried down with the tangled mass of steel. In only a few seconds the destruction was complete. Twenty thousand tons of steelwork were lost in the crash, making it one of the most startling catastrophes in bridge engineering history.

This greatly affected Theodore Cooper who, after a lifetime of contributions to the art of bridge building, was then at the peak of his professional fame. When word reached him early that day that the outer end of the cantilever showed excessive deflection, he frantically dashed off a telegram ordering every man off the span, but the message arrived too late. Following the disaster, he retired in seclusion and a few years later he died a broken man.

The investigation which followed disclosed that the collapse was caused by the buckling of a compression member that had been inadequately braced. Previously accepted empirical rules for the design of sections and details of compression members, tried and tested for smaller members, had betrayed the designer when they were applied to compression members of larger and unprecedented dimensions. As a result of the large-scale experiments and studies that followed, the design and detailing of large compression members were brought to a scientific basis. In addition, attention was now directed to the proper design and construction of the joints between compression members and to the analysis and elimination of "secondary" stresses produced in truss members by their deformations. The Quebec disaster of 1907, more than any other occurrence in the evolution of bridge building, revolutionized the art by bringing it to a new high level of scientific analysis and design.

Up to 82 men were carried to their death in the tangled mass of steel when the Quebec Cantilever Bridge collapsed in 1907. Also lost in the crash was 20,000 tons of steel work. — JIM SHAUGNESSY COLLECTION

With a new design of greater rigidity, Canadian authorities decided to try again, this time not using the cantilever method of erecting the center span. Instead of building out from the cantilever arms, the lifting method was adopted. By means of a system of powerful hydraulic jacks, the span, weighing 5,100 tons, was to be raised from barges to a total height of 150 feet.

In Sillery Cove, a little town some three miles downstream, the suspended portion of the bridge was built on falsework over barges. On full tide, early in the morning of September 11, 1916, the barges left the Cove with their precious load and by 7:40, were centered between the two great steel arms. Notch by notch, with careful pumping of the hydraulic jacks, the span was slowly lifted from its supporting barges. Everything was proceeding well.

At 10:50 A.M. there was a sharp report and something seemed to snap. The 5,100-ton span, which had been raised about 12 feet, suddenly slipped out of its stirrups and went crashing into the river, carrying eleven men to their death and costing the bridge company a million dollars.

A careful examination of the cantilever arms revealed that they had not been damaged by the fall, despite the resulting

The center span of the Quebec Cantilever Bridge was floated upstream early in the morning of September 11, 1916. The 5,000-ton span was then lifted into place notch by notch. It had been raised about 12 feet, with everything going well, when suddenly something snapped and it went crashing into the river. — BOTH JIM SHAUGHNESSY COLLECTION

vibrations, when the 5,100 tons were suddenly released.

It was decided to proceed with the reconstruction of the span with all speed, since there was now a war in Europe and the bridge was urgently needed to carry men, ammunition and an increasing amount of freight.

By the end of August 1917, the new span was completed at Sillery Cove. Again on barges, on the highest tide of the year, the new center span left the Cove and proceeded up river. Immense crowds watched the span being moved into position. The hoisting gear was made fast by 7:00 A.M. and at 9:10, the first lift was made. During the opening stages of the task, mindful of the previous year's calamity, the workmen were understandably nervous. Their confidence returned after the first two or three lifts, and the work proceeded more rapidly.

The engineers, however, were very cautious by now and took no risks. The lifting operation, scheduled to take 30 hours, was suspended for the first day after 12 lifts had been made, and the span was left that night about 30 feet above the water. The following

At long last, on September 20, 1917, the center span was lifted into place. The bridge was secured at all points by 4:00 P.M., leaving only minor connections to be made. — JIM SHAUGHNESSY COLLECTION

In this recent photograph of the 3,239 foot Quebec Cantilever Bridge, one can see the new highway bridge which was built to supplement it. This big bridge still fulfills a vital role in carrying rail traffic as well as the vehicular traffic — an essential link between the province of Quebec and the Maritime provinces to the east. — RICHARD J. COOK

day it was satisfactorily passed through another 22 lifts, or 44 feet, while on the third day 26 lifts were recorded, leaving the span within 30 feet of its intended permanent position.

Finally, on the fourth day, at 4:00 P.M., September 20, 1917, the bridge was secured at all points, leaving only minor connections to be made.

Locomotive cranes laid the floor, and concrete sidewalks were poured. By October 17, the track was laid and the first train passed from bank to bank. On December 3, the track was turned over to the operating department of the Canadian Government Railways and the first regular freight train crossed the span.

A newspaper account recorded the event stating:

> It is 11 o'clock in the morning, December 3, 1917. George Walker and Edmund Parsons, two of the oldest employees of the Transcontinental (predecessor road of the Canadian Government Railway — later Canadian National), stationed their steam locomotives at the north approach of the Quebec Bridge.
>
> While they got down from their cab, the guests gathered around the engine for a photograph. The locomotive carries number 2900. In the group present, one can see some people wearing woolen capes, others are in overalls. One gentleman with a derby hat stands on the cowcatcher. The moment is historic. The first regular freight train is about to cross the St. Lawrence on the Quebec Bridge.
>
> No official ceremony accompanied this event which is, however, one of great importance for our city and even for the country. It is believed that there hasn't been any official ceremony since the passenger trains have started crossing the bridge for the last fortnight.
>
> Several government railway officials came to Quebec from Moncton for the purpose of crossing the bridge in a private car coupled into a freight train.

Among the witnesses of this event was Alcide Martineau. Fifty years later he recounted in a newspaper interview: "I remember that I saw the first run made when the railway released the first passenger train. Trains crammed with steel had crossed the bridge to prove its solidity."

A soldier stands guard on one of the piers of the Quebec Cantilever Bridge during World War I. This photograph gives an excellent comparison of its size. — JIM SHAUGHNESSY COLLECTION

The newspaper account stated, "He also remembered the visit of Henry Ford with his four-cylinder car which had its wheels temporarily replaced with flanged wheels so that it could run on the rails. The passing of the first regular freight over the Quebec Bridge wasn't given much publicity. Everyone was more occupied with the Laurier election than this historic event. Canada was at war and the politicians were making life hard."

The 3,239-foot Quebec Bridge ranked as the longest and largest of its kind, but was one of the most graceful structures wrought in metal. Its construction involved the setting of 66,480 tons of steel. Few bridges have been carried to a successful completion in the face of such difficulties, and few have survived such disasters. It stands as an enduring monument to the faith and skill of the engineers.

Today the Quebec Bridge, which also carries a motor roadway, designed to carry 2.3 times the load designed for the Forth Bridge in Scotland, forms an essential link between the Province de Quebec and the Maritime Provinces with their important eastern ports.

One of the great American cantilever bridges is the Pittsburgh & Lake Erie Railroad bridge spanning the Ohio River between Beaver and Monaca, Pennsylvania. The first train ran over this bridge on May 11, 1910, replacing the small bridge at the left. — DONALD DUKE COLLECTION

Other great cantilever bridges have also figured in the railroad scene. Among them are the Pittsburgh & Lake Erie bridge at Beaver, Pennsylvania; the Michigan Central bridge over the Niagara River at Niagara Falls, Ontario-New York; the Pittsburgh & West Virginia (now Norfolk & Western) bridge over the Ohio River at Mingo Junction, Ohio; The Wabash (now Norfolk & Western) bridge over the Missouri River at St. Charles, Missouri; the Cotton Belt (St. Louis Southwestern) bridge over the Mississippi at Thebes, Illinois; the Illinois Central bridge over the Mississippi at Vicksburg, Mississippi; the Missouri Pacific bridge over the Mississippi at Baton Rouge, Louisiana, and the Huey Long Bridge over the lower Mississippi River at New Orleans. The Vicksburg, Baton Rouge and the Huey Long bridges also carry highway traffic. The Huey Long bridge, 4.4 miles long (with approaches) is said to be the longest bridge in the world.

Albert Lucius, born in Germany, opened his office in New York City in 1886. It was he who was chosen to be the engineer for the great cantilever bridge which was to span the Ohio River at Beaver, Pennsylvania, on the Pittsburgh & Lake Erie Railroad. Influenced by the 1907 Quebec Bridge disaster, the McClintic-Marshall Co. adopted a procedure that placed joint responsibility on the bridge designer, fabricators of the bridge members and the erectors of the structure. Albert Lucius, consulting engineer, was to submit his design of each bridge member to the McClintic-Marshall Co., the fabricators and the erectors, and they were not to proceed unless and until the design was approved. This careful control helped to account for the great strength of the bridge, designed to carry trains of iron ore, coal and finished steel products, as well as loads of general freight. The cantilever at Beaver was hailed as one of the great bridges of the world, even though there was an extensive use of eyebars and pin connections rather than the more contemporary method of riveting — and thought by many to be superior.

Prior to regular use of the bridge, a test train of ten of the road's heaviest locomotives, each weighing 336,000 pounds with tender, and 20 freight cars loaded with iron ore and each weighing 150,000 pounds, were placed on the bridge. Compression measurements, taken under the 2,842-ton load and also taken afterwards when the load was removed, showed that the bridge met more than was required. On May 14, 1910, passenger train

After passing over the Ohio River Bridge, a steam locomotive moves a coal train through Beaver station.

No. 5 was the first regular train to use the bridge.

The P&LE bridge replaced a single-track pin-connected structure built about 1890. It measures 1,787 feet long, center to center of end pins or about 1,880 feet long over all of the steel work. There is one approach span of 320 feet. The channel span, measuring 769 feet, is made up of two 242-foot cantilever arms and a 285-foot suspended span. The 16,000-ton mass of steel cost about $1,300,000 to erect, plus an additional $700,000 for the approaches. The bridge is located at the mouth of the Beaver River where the Ohio River turns in a westerly direction.

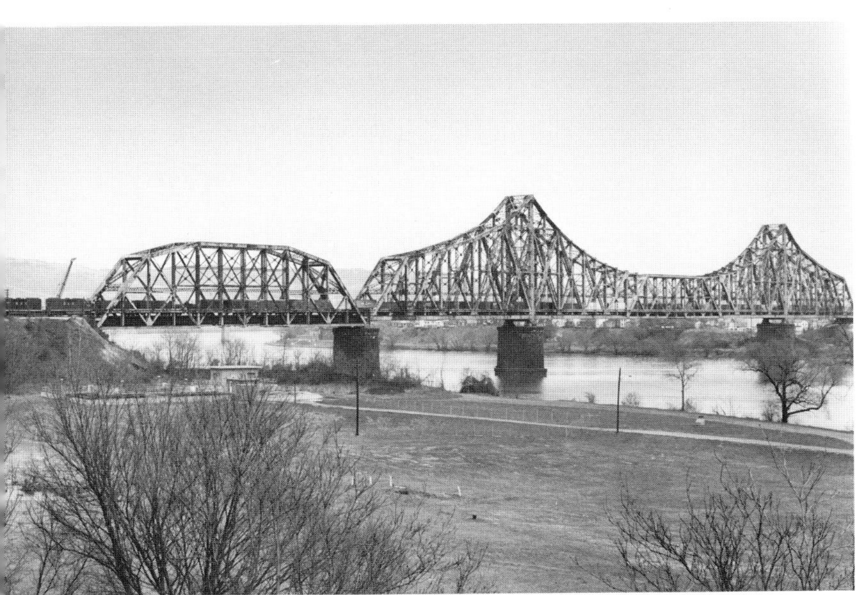

At the upper left, a towboat moves down the Ohio River and passes under the P&LE Ohio River Bridge. This view is taken from the Monaca side of the river. (LEFT) The P&LE bridge not only carries its own trains, but also those of the Baltimore & Ohio. Here a B&O freight rushes toward Chicago. — BOTH RICHARD J. COOK

This drawing shows the New York Central's cantilever bridge over the Niagara River Gorge. The 1,910-foot bridge was officially opened to traffic on December 20, 1883. — SMITHSONIAN INSTITUTION

One of the first cantilever bridges to be built in North America was the one engineered by Charles C. Schneider for William H. Vanderbilt's Canada Southern, a subsidiary of the New York Central. When Vanderbilt found that the Great Western refused him the use of its bridge over the Niagara Gorge, he decided to have his own built a few hundred feet upstream. Work began on April 15, 1883 and the 910-foot long bridge was officially opened to traffic on December 20, 1883, and remained in service until 1925, when it was replaced by the steel arch bridge which is still in use.

Train movements over the Missouri River at St. Charles, Missouri, were greatly improved on October 29, 1936 when a new cantilever bridge of the Wabash Railway was completed and opened for traffic. The new bridge replaced an 1868 bridge and was then again extensively rebuilt in the early 1880's. The new bridge, moved to a different location, together with line revisions and approaches, constituted a new 'route' of seven miles in length. As in the case of other railways, seeking to replace older structures over large navigable rivers, the Wabash was compelled to provide a new bridge that embodied span lengths which bore no relation to the spans of the bridge to be abandoned. The longest span on the old bridge was 321 feet; but the bridge that replaced it has a channel span of 600 feet, clear width, or 624 feet, center to center of piers. This span comprises the central unit of a three-span cantilever structure that is distinctive by reason of the marked difference in length of the anchor arms, 274 feet and 431 feet respectively. A fourth span, a simple span type, is of the same length, center to center, 312 feet, as the suspended span of the cantilever structure. These four spans, comprising the river structure, have a total length of 1,645 feet, center to center of piers. The 92-foot high superstructure provides a clear headroom of 45.5 feet above the high water level of 1903, which was 26.5 feet above standard low water.

The Wabash Railroad built a new cantilever bridge over the Missouri River at St. Charles, Missouri, in 1936. The new bridge with its long approaches replaced an outmoded 1868 era bridge. This bridge now forms a part of the Norfolk Southern System. — RICHARD J. COOK

Cotton Belt's cantilever bridge at Thebes, Illinois, built in 1905, is an important rail link. Not only is it the crossing of the Mississippi River for the Cotton Belt, but it is a shortcut route for the road's owner, Southern Pacific, in its Los Angeles to St. Louis fast freight route. — RICHARD J. COOK

Southern Pacific's *Prosperity Special* was a train of 20 2-10-2 type locomotives all coupled together en route from the Baldwin Locomotive Works at Eddystone, Pennsylvania, to the SP in California. This 1921 event was to show that good times were coming when the nation's economy was down. The SP's order was for 50 of these 2-10-2 locomotives. In this scene, the *Prosperity Special* was photographed crossing the Cotton Belt's (St. Louis Southwestern) cantilever bridge at Thebes. — DONALD DUKE COLLECTION

The turn of the century saw plans proceeding for yet another cantilever bridge, this time over the Mississippi River at Thebes, Illinois. Engineers Alfred Noble and Ralph Modjeski sought the elimination of the transfer boats used in the transfer of trains between the Illinois Central, the Chicago & Eastern Illinois, and the St. Louis, Iron Mountain & Southern railroads on the east bank, and the Frisco System, the St. Louis, Iron Mountain & Southern, and the St. Louis Southwestern railroads on the west bank. At this point the river is about 2,700 feet wide between high water banks, and 2,400 feet wide between low water banks, and has a depth of about 50 feet, the level being subject fo floods in the Ohio, Missouri and upper Mississippi.

The Thebes Bridge opened for traffic in 1905. At the time it was one of the most notable bridges in the United States and, in addition to its impressive steel work, it boasts of having two arched concrete approach viaducts. The one on the east side has five semicircular arch spans of 65 feet, and the third pier from the river forms an "abutment" pier of larger size than the others. The viaduct on the west side has six spans of 65 feet and one of 100 feet. The bridge has spans of 518.5 feet, 521, 671, 521 and 518.5 feet; 2749.10 feet overall.

Early in 1922 when business was slow in the United States and a return to greater industrial activity seemed slight, the Baldwin Locomotive Works wanted to create a sense that prosperous things were happening, so they devised a public relations scheme to bring attention to the fact that they were making a large delivery of locomotives to the Southern Pacific. In the delivery, they coupled 20 freight locomotives together all at one time and called the movement from Eddystone, Pennsylvania, to the Southern Pacific lines the "Prosperity Special." The train of locomotives, each weighing 441,900 pounds, was moved over Cotton Belt rails in a solid train from East St. Louis, Illinois, to Corsicana, Texas. And of significance is the fact that the Thebes Bridge over the Mississippi could handle such a heavy movement.

On account of railroad rivalries during the railroad building era, two significant cantilever bridges instead of one were built over the Mississippi River at Memphis, Tennessee.

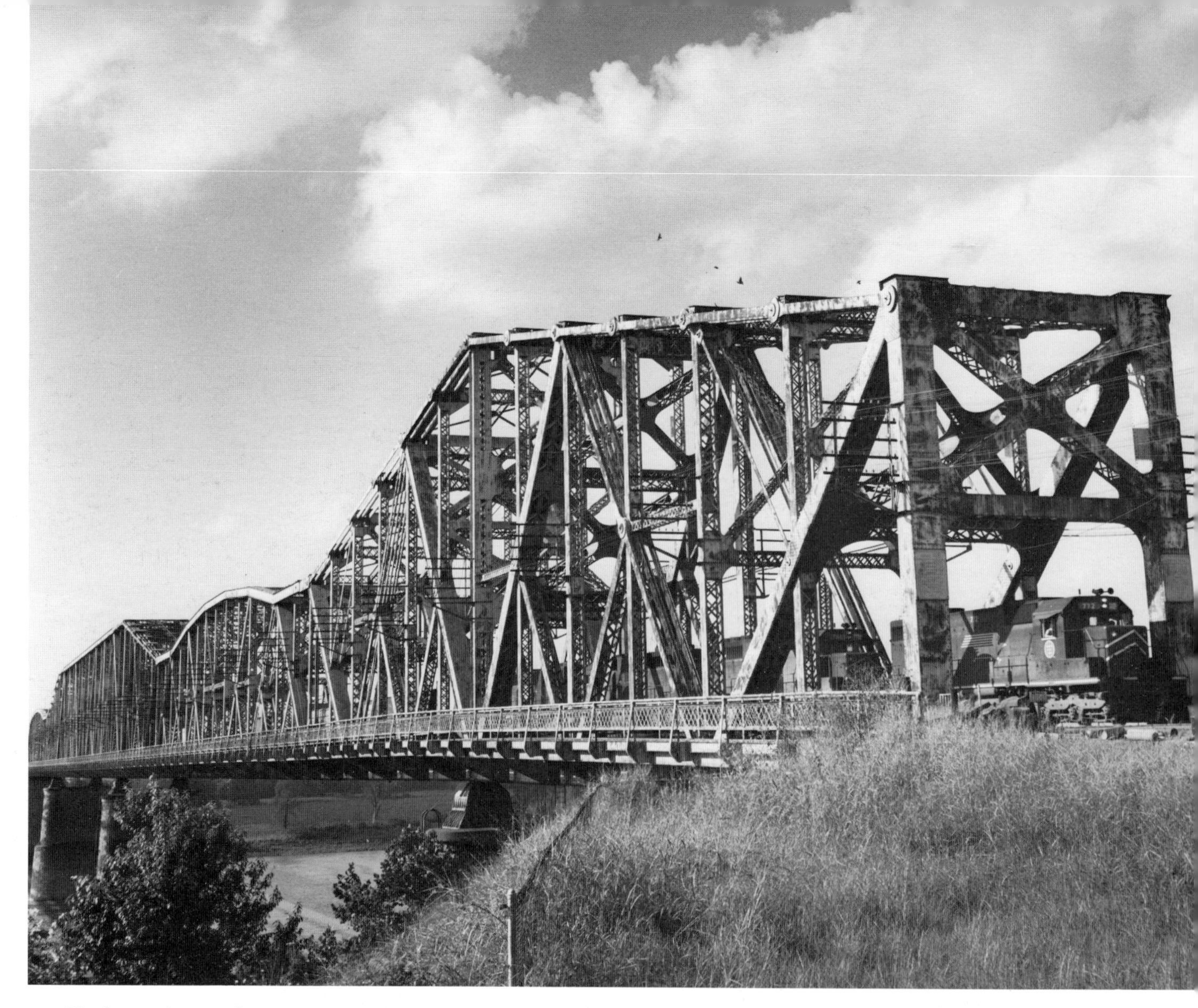

Today, in order to save money, a joint operation over a single structure would have made more sense.

The first structure at Memphis, called the Memphis bridge, was completed in 1892. Its construction was under the able hands of engineers George S. Morison and Alfred Noble. The structure is an irregular combination of cantilevers, anchors and suspended spans, a pattern dictated by the position of the channel and the shore's topography. The west approach, running over a low-lying, easily flooded area on the Arkansas side of the river, is an impressive mile-long steel viaduct of Warren deck trusses. Built by the Kansas City, Fort Scott & Memphis Railroad (better known as the St. Louis-San Francisco Railroad), the bridge itself combines a 790-foot cantilever span with two 621-foot continuous trusses, one on either side, and carries a public highway over the river as well

The Harahan Bridge over the Mississippi River at Memphis, Tennessee, was built by the Chicago, Rock Island & Pacific Railroad in 1911. This photograph shows a Missouri Pacific freight crossing the bridge. — RICHARD J. COOK

as the single track line of railway.

The second Mississippi River rail crossing is the Harahan bridge, alongside the Memphis bridge. In 1911, the Chicago, Rock Island & Pacific Railroad decided to build a double-track bridge over the Mississippi in order to compete with traffic coming over on the Frisco's Memphis bridge line. J.T. Harahan, retired president of the Illinois Central Railroad, was elected president of the newly formed Arkansas & Memphis Railway Bridge & Terminal Company. The proposed bridge was planned to accommodate the trains of the Cotton Belt Route, and an automobile highway.

The Harahan bridge rests on six piers and has two cantilever spans, one of 790 feet (the channel span) and a shore span of 604 feet. The center, the fixed span, is 621 feet in length. The total length of the main bridge, counting between centers of extreme end pins of the spans is 2,548 feet and 10½ inches. The Arkansas approach viaduct measures 2,364 feet in length. Ralph Modjeski was the consulting engineer in charge of the bridge project.

On July 15, 1916, the first trainload of rail freight crossed the newly completed bridge, even though the vehicular roadways were only about half completed. The first automobiles did not cross the bridge until August 31, 1917, and the formal grand opening of the structure was not held until September 5, 1917.

Following the dedication, a series of misfortunes occurred involving the bridge. The first came on April 29, 1928, when fire destroyed a 225-foot section of the approach viaduct on the Arkansas side of the river, forcing auto traffic to travel over muddy river bottom roads in order to reach its destination. Again, on September 17, 1928, an 800-foot section of the Harahan bridge was rendered useless after a fire that, for a while, threatened to destroy the whole structure. Rail traffic resumed on November 15, 1928, only to be interrupted by a second fire eight hours later. Fortunately, this fire was not serious.

Another Missouri Pacific (now part of Union Pacific) fast freight rumbles into Memphis over the Harahan Bridge. — RICHARD J. COOK

In the above scene, a freight train is framed by the gigantic steelwork of the Harahan Bridge. (LEFT) The Memphis rail crossings of the Mississippi River from the Arkansas side of the river. At the far left a Louisville & Nashville freight starts across the Harahan Bridge. The bridge at the right belongs to the St. Louis-San Francisco Railway (the Frisco). — BOTH RICHARD J. COOK

The imposing 4.35-mile double-track Huey P. Long-Mississippi River Bridge rises 135 feet above the water level, and the main water crossing itself is 3,525 feet long. Trains approach this bridge on a 1.25 percent grade on each side, with a west approach of 10,790 feet, and an east approach of 8,680 feet. The freight train on the structure is hardly visible in this photograph, somehow befitting the majesty of this great bridge. — RICHARD J. COOK

Ralph Modjeski, a specialist in designing cantilever bridges over the Mississippi River, teamed up with Masters and Case to form the firm of Modjeski, Masters & Case, Inc. Their first project was a proposed bridge over the Mississippi River at New Orleans where the river is almost two-thirds of a mile wide and very busy with ocean-going water traffic. The Louisiana legislature gave the go-ahead in 1932 to build one of the largest and most costly bridges in the world up to that date. The need was great and in 1933 work began on the Huey Long Bridge. New Orleans' commercial growth depended on increased rail traffic and better rail access to growing dock areas.

The bridge was completed in 1935 and provided for two tracks plus a roadway on either side of the tracks. The bridge, located 3.3 miles west of the New Orleans business center, provides a 135-foot clearance above high water level and is characterized by its very long approaches. The east approach is 8,680 feet long; the west approach is 10,791 feet long. Approach grades are 1.25 percent and result in the very slow movement of trains. Amtrak's *Sunset Limited* uses this bridge and provides a thrill to those passengers who can imagine themselves "flying" as they look down and see treetops and houses well below.

The length of the main river crossing is 3,525 feet. It consists of, from the east bank to the west bank, three 267-foot deck truss spans, one 330-foot deck truss span, one 528-foot simple through truss span and a three-span through truss cantilever structure made up of a 790-foot central span and two 528-foot flanking spans. Trains of the Southern Pacific (Texas & New Orleans) Missouri Pacific and Illinois Central Gulf still use the bridge.

As the sun sets in the west, the *Sunset Limited* crosses the Huey Long Bridge for a 6:00 P.M. arrival in New Orleans. — DONALD HOFSOMMER COLLECTION

Prior to the Huey Long Bridge, Southern Pacific (Texas & New Orleans) trains reached New Orleans by car ferry across the Mississippi River at Avondale. — DONALD HOFSOMMER COLLECTION

A military camp was built below the Huey Long Bridge during World War II. — DONALD HOFSOMMER COLLECTION

The first regularly scheduled train to use the $13 million Huey Long Bridge was the eastbound *Sunset Limited* on December 17, 1935. The jungle at the foot of the bridge, at the left, became a military camp during World War II. — DONALD HOFSOMMER COLLECTION

Poughkeepsie Bridge of the Central New England

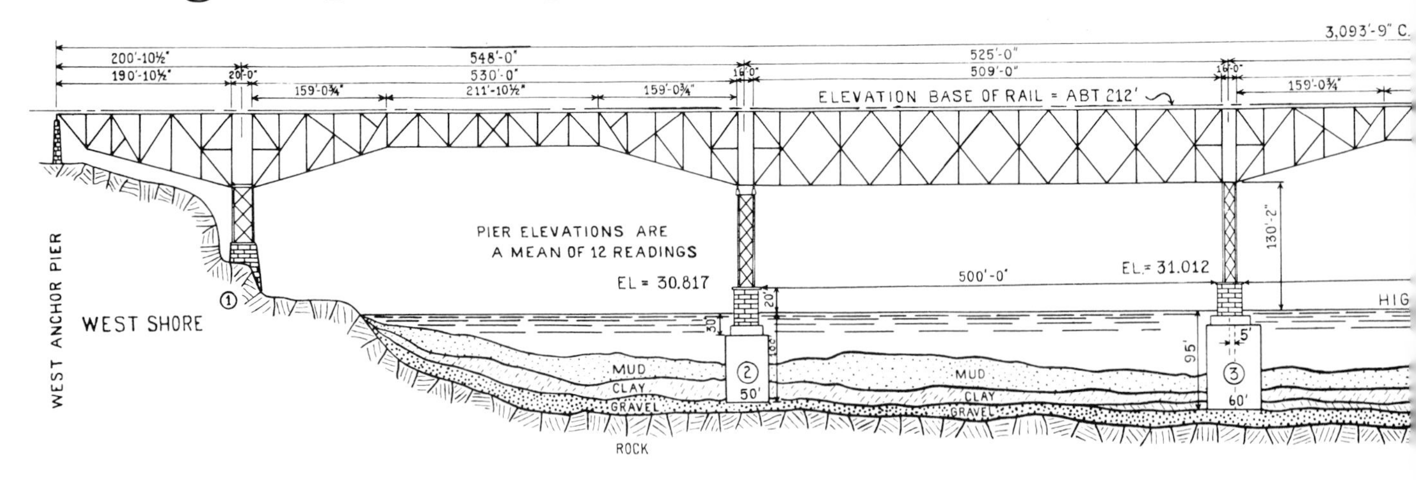

Another cantilever bridge of note is the ill-fated Poughkeepsie Bridge at Poughkeepsie, New York. No longer in use but still standing, this bridge over the Hudson River was built early on: 1888! It linked New England with the eastern Pennsylvania coal fields, the west shore of the Hudson River, and its connection to New York City and cities in New Jersey.

Organized in 1871, the Poughkeepsie Bridge Co. decided to use the newly popular cantilever construction. Although the bridge was completed in December 1888, the first paying train, a Barnum circus train of 40 cars, did not pass over the bridge until May of 1889. At that time there were two tracks on the bridge. During a move to strengthen the bridge in 1906-1907, the double-track was overlapped, or gauntleted, so that the bridge could accommodate larger 2-10-2 locomotives. The total length of the bridge, with its approaches, is 6,768 feet. There are two cantilever spans of 548 feet, two connecting spans of 525 feet, and two anchor spans of 201 feet each. One of the highest bridges in the East, the Poughkeepsie bridge rises 212 feet from the water below to the base of the rail. Regular passenger service over the structure ended in 1927 and the bridge, since that time had been used exclusively as an important link in the New York, New Haven & Hartford's freight line. Since Penn Central, then Conrail came to own the New Haven route, the bridge, badly in need of repair and suffering from fire damage, was declared "excess" by the latter system. Now trains have to travel many unnecessary miles in order to reach New Haven or Boston.

The majesty of the Poughkeepsie Bridge can't be realized until the viewer stands on the banks of the Hudson River and looks up at this huge cantilever bridge. A bridge across the Hudson River at Poughkeepsie had long been dreamed of as a short-cut rail route from New England to the coal fields and the great expanse west of the Hudson. Until the bridge was built, rail traffic was ferried across the river from Newburgh to Fishkill. The Poughkeepsie Bridge Company was organized in 1871 to build the bridge. At first a suspension bridge was proposed, but proved impractical. A cantilever bridge was built and opened in 1888 and was strengthened in 1906 to accommodate larger steam locomotives. With the formation of Penn-Central, the bridge was bypassed as a through-route, and eventually abandoned. The bridge is now owned by Conrail and many attempts have been made to have the bridge repaired as several years ago a few ties and deck timbers burned at the east end of the structure. This magnificent cantilever bridge, shown at the right, now stands as a silent monument to railroading of a by-gone era. — RICHARD J. COOK

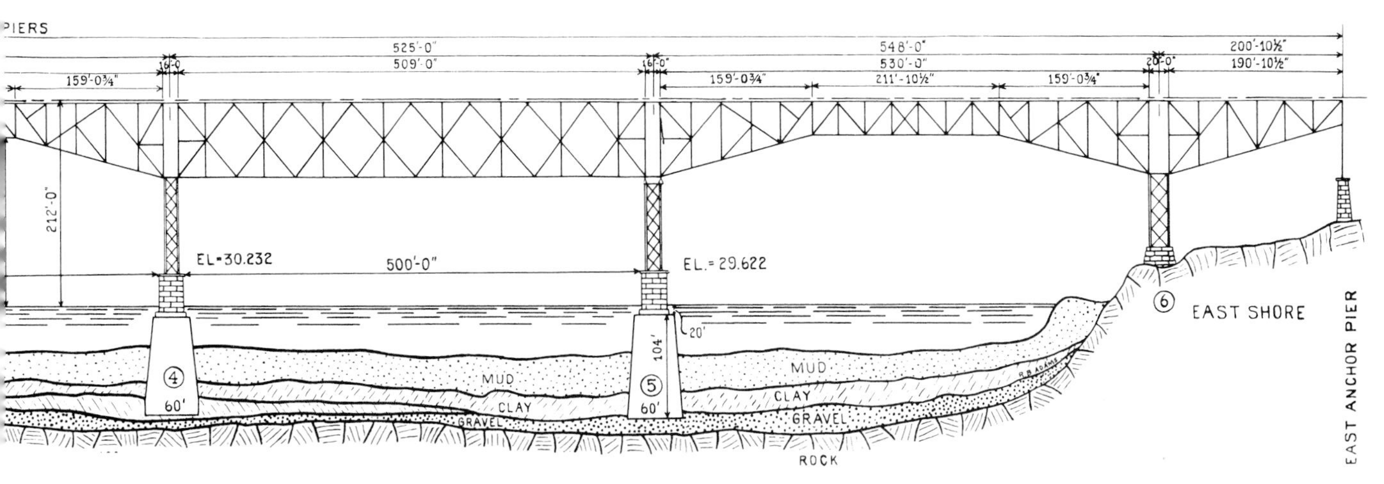

ROBERT B. ADAMS — COURTESY OF NATIONAL MODEL RAILROAD ASSOCIATION BULLETIN

Erecting the Old Pecos River "High Bridge"

"Highest in the World," they said of the Pecos River "High Bridge," this spidery-looking structure in West Texas. Here, in the construction stages, may be seen the traveller crane which was used to place each girder beyond the last one and then anchor it into place on the steel piers below. — BAKER TEXAS HISTORY CENTER

At the left, the "High Bridge" is just about finished. (RIGHT) A Southern Pacific passenger train, perhaps one of the first to cross the bridge, negotiates its way over the deep canyon below. — BAKER TEXAS HISTORY CENTER

Today's Pecos River bridge towers over a barren area of south Texas where, aside from the rumble of a passing train, the only sounds are the lapping of the waters far below or the occasional cry of a bird passing overhead. The present bridge utilizes a continuous cantilever deck, resting on hollow concrete piers and rising up to 275 feet. The bridge, with an overall length of 1,390 feet, was designed and the construction supervised by Modjeski and Masters of Harrisburg, Pennsylvania. — RICHARD J. COOK

The "High Bridge" was strengthened in 1910 when another line of girders, supported on new cross girders, was placed between the legs of the towers. At the west end, 19 spans were removed and replaced by a filled embankment. — SOUTHERN PACIFIC COLLECTION

No story of the railroad truss bridge in North America would be complete without including one of the highest bridges in the world, or perhaps it should be said: of the *two* highest bridges (the very highest can be a matter of some dispute). A few miles from the point where it flows into the Rio Grande, the Pecos River passes through a canyon which is directly on the route of the Southern Pacific (Texas and New Orleans) line.

The first Pecos River Bridge was begun in 1891 and completed in 1892. It carried a single-track and, at the time, was one of the longest bridges in the United States. It was originally constructed in 46 spans; 32 of these spans were of 35 feet, two of 35 feet 3 inches, one of 45 feet and eight of 65 feet. There were two pin-connected cantilever trusses and anchor arms of 172 feet 6 inches, and one 80-foot suspended span. The overall dimensions, including approaches, were 2,180 feet. When the new bridge opened, it shortened the distance between San Antonio and El Paso, Texas, by 11 miles, besides saving heavy grades and avoiding many tight curves.

The erection of this spectacular bridge was exciting in itself. A nine-ton traveller crane, designed to run on the newly laid rails, was used to move back and forth to build the towers and girders until the cantilever trusses were reached. Then this part of the erection was carried on by building out from each side until the center was reached and the members joined at last to complete one of the world's highest bridges.

The original (1892) "High Bridge" of the Southern Pacific (Texas & New Orleans), spanned the Pecos River in south Texas. At the time, it was reputed to be the highest bridge in North America. — SOUTHERN PACIFIC COLLECTION

The "High Bridge," as it was called, was strengthened in 1910 by the addition of another line of girders that were supported on new cross girders between the legs of the towers and, at the west end, 19 spans were removed and replaced by a filled embankment.

It was one of the wonders of the railroad engineering world when it was completed and a highlight for passengers who were crossing West Texas on the Southern Pacific's passenger trains. An article about the bridge, appearing in *Railway Progress* in 1952, tells of the lore and wonder associated with Texas's "High Bridge."

> Three hundred twenty feet above the snarling rapids of the Pecos River, daring workmen moved cautiously along a strange looking structure that was to become famous as a miracle of railroad engineering. It was 1891, and determined railroad men had devised unique but crude machinery in an attempt to build a bridge that would eliminate eleven miles and two long tunnels from the Southern Pacific's line linking important cities on the Gulf Coast with those on the Pacific.
>
> Suddenly and without warning, the Texas air was pierced by a grinding crunch of metal, the screams of men and the crash of steel girders into the river bed. A section of the partly completed structure, one of the world's first steel cantilever spans, had fallen and sent ten men plunging into the bottom of the canyon. Seven were killed instantly and three were critically injured.
>
> A call went out for Judge Roy Bean, the "Law West of the Pecos," and a man already famous for such rulings as his "I can't find any law against killing a Chinaman." Judge Bean came from his judicial headquarters in his saloon at nearby Langtry to conduct an inquest. It was a task the judge was to perform several times during the two years the famed Pecos River High Bridge was under construction.
>
> The bodies of the seven men who had been killed were laid in a row. Judge Bean looked carefully at each one of them and then at the wreckage. Over each body he said: "This man seems to have come to his death by the big timbers falling on him." The judge then turned his attention to the three critically injured men and, although they were still alive, he pronounced the same verdict over each of them. The coroner's jury had been selected from the onlookers, most of whom had been working on the bridge. One member of the panel after hearing the verdict, had the courage to speak up.
>
> "These men aren't dead, Judge," he said.
>
> "That's all right," Judge Bean retorted, "they soon will be and I'm not going to ride up here again just to hold another inquest."
>
> Thus, almost from its inception, grisly humor was mixed with tragedy in a manner that was to become a familiar pattern for the bridge, once the world's highest. During its more than half century of service, the Pecos River High Bridge was the scene of such varied adventures as one of the most daring stunts in aviation history, a disappointed lover's leap to death, several weddings and two death-defying horseback rides by a woman. It sheltered, too, for almost a quarter of a century, a former United States marshall who wanted to get away from it all.
>
> Approaching trains had to come to a dead stop for about ten minutes before crossing it to let the swaying of the bridge, caused by the vibrations of the rails, to stop. Even then, the top speed limit was just twelve miles per hour on the bridge. The stops gave thousands of passengers a chance to get off the train and admire the famous structure.
>
> Mere sight of the bridge terrified Mike O'Reilly, a Southern Pacific fireman who often boasted he could whip any man twice his size. He admitted his fear of the bridge although he made it clear that he was afraid of nothing else. He never made a trip across the structure without audibly heaving a sigh

On June 24, 1922, the *Prosperity Special* crossed the Pecos River "High Bridge" heading west. One can easily imagine that this train of heavy steam locomotives would move very slowly over the old bridge. — LIBRARY OF CONGRESS

of relief and declaring "the saints in Heaven were with me," for getting him across safely.

A young woman daringly rode her horse across the bridge not long after it was completed. Miss Patty Moorhead (now Mrs. E.P. Bell of Del Rio) was out riding with a neighboring young rancher one day when she challenged him to ride their horses across the bridge, using the board walkway for a passage. Astride their horses, which were accustomed to any terrain if not to crossing a river 321 feet above the water, they rode across just for the thrill.

Later, Miss Moorhead repeated the stunt as a matter of convenience. Late one afternoon, she found herself far from home without enough time to get there before dark if she followed the usual route around the deepest part of the canyon. So, once again, she guided her horse across the famed bridge.

The bridge was guarded at various times by the Texas Rangers, the National Guard and several different units of the United States Army. During World War II, the bridge was under constant surveillance by companies of the 766th Military Police Battalion. They patrolled the structure relentlessly to prevent its being damaged by saboteurs, for it played an important role in transportation of vital war materials.

Its most famous guard was J.R. Hutchins, whose adventures as a law enforcement officer soured him on the world. Said to be the youngest United States marshall in history, he got his badge and an assignment in the Oklahoma Indian Territory at the age of eighteen. He took a job with the Southern Pacific as a guard and inspector in 1924 and moved into the small cottage beneath the bridge.

"I grew tired of shooting at my fellow man," Hutchins said, "or seeing him punished for a crime, even though he deserved it. I grew sick of double-crossing, seeing men's souls bartered for selfish gain and having to live an existence of intrigue. I craved honesty in all dealings, regardless of what they might be, and I saw I couldn't attain my desires as a law enforcement officer."

Hutchins seldom ventured out of the canyon. Train crews lowered groceries and other necessities to him by rope for many years. He had a radio in his cottage and is said to have told his friends he had never heard anything on the radio which made him want to move back to the outside world.

An Army unit was guarding the bridge one June night in 1917 when an attractive young lady, who had a ticket from Del Rio to Sanderson, joined other passengers who got off their train when it stopped at the bridge. They walked to the edge of the gorge to look at the famous structure.

She left the other passengers and was challenged by one of the soldiers guarding it. She was permitted to pass when she explained that she only wanted to see the river. Suddenly, she jumped off the bridge and plunged into the canyon. She had left her shoes on the train and inside one of them was a note explaining that she had been disappointed in love.

Shortly after World War I ended, Jimmy Doolittle was a member of an air squadron which patrolled the Mexican border. He flew over the bridge frequently and told some of his friends several times that some day he was going to fly under it.

With the same daring he was to exhibit in his famous Tokyo raid more than two decades later, he did fly under the structure in March, 1921. He had to tip the wings of his open cockpit plane in order to clear the vertical piers of the bridge.

When World War II brought a sudden increase in the volume of freight, SP engineers took another detailed look at the structure which stood as a testimony to the courage and ingenuity of their profession's pioneers. They decided the increased loads and more frequent crossings might make it impossible to keep up with the repairs necessary to maintain the safety of the bridge.

Because of the importance of the transcontinental line, they decided that the old bridge must be replaced by a modern one fully capable of carrying the freight of an America that had gone to war.

The new bridge, 440 feet downstream from the old location, was put into service on December 21, 1944, when the SP's "Sunset Limited" passed over it for the first time.

The original High Bridge, by then dear to the hearts of many people who lived in the area, had successfully weathered one World War and most of another in addition to its long years of peacetime service.

Construction of such a bridge today would be comparatively easy, thanks to modern equipment and present day methods. But the building of it under conditions existing sixty years ago long will be remembered as a marvel of engineering.

The present Pecos River Bridge, which carries a heavy amount of freight between California, Texas, and the South, stands majestically under the warm Texas sun. This bridge is located between Sanderson and Del Rio, in western Texas, where the desert topography is only interrupted by the Pecos River crossing. Thus is required this spectacular — if isolated — feat of engineering. — RICHARD J. COOK

A construction company bought the original High Bridge and announced plans to dismantle and move it to Guatemala, where it was to be reassembled over another gorge and used as a highway bridge. The Guatemala deal fell through however, and the steel was sold for scrap.

Never to be scrapped, however, are the memories the Pecos River High Bridge left behind and its fame as one of the world's most remarkable engineering feats.

The present bridge utilizes a continuous cantilever deck on hollow concrete piers rising to 275 feet. The new bridge, much more massive than the old spidery-looking "High Bridge," is designed for much heavier loads, and also compensates for earthquake forces. The rock cliffs at the site are sound and about 770 feet apart, while the gorge is 322 feet deep from the base of the rail to the water level below. Limestone rock is the founding material for all piers. Overall length of the bridge is 1,390 feet. The bridge was designed and the construction supervised by Modjeski and Masters of Harrisburg, Pennsylvania.

Except when a Southern Pacific freight or the Amtrak *Sunset Limited* passes over the bridge, the surrounding countryside is exceptionally quiet, with only the mighty bridge dominating the visual landscape. In the hush that follows after the rumble of a passing train, the only noises that can be heard are those of the metallic snapping as the bridge metal expands in the warm Texas sun, the cry of a passing bird and the gentle sound of the Pecos River waters lapping against the rocky shore some 300 feet below. The quiet is as awesome as the sight of this mighty mass of steel and concrete.

Another truss bridge of note, built in a more modern time, is the Chicago, Rock Island & Pacific bridge over the Cimarron River at Liberal, Kansas. This structure, an integral part of the Arkalon Cut-off, was built in 1938-39 and opened in July of 1939. Because of its husky appearance, it was soon given the nickname of "Sampson of the Cimarron." It is 1,269 feet long and consists of five single-truss 250-foot deck spans resting on reinforced concrete piers and abutments. Although the river looks passive most of the time, it is known for its temperamental character at floodtime. To protect the long fill approaches to the bridge, floodwater dikes, built upstream, were surfaced with sheet steel piling and concrete slabs. The bridge replaced two pile trestles and eliminated the need for an additional 12 miles of difficult track in order to cross the river.

The bridge, although no longer belonging to the defunct Rock Island, is still in use, now as an integral part of Southern Pacific's Cotton Belt "Golden State Route" to Kansas City from El Paso.

The Rock Island's main line southwest to New Mexico and the Southern Pacific connection at Tucumcari, crossed the Cimarron River east of Hayne, Kansas. "Samson of the Cimarron" is a tremendous steel and concrete structure of great dimensions and beauty. Today the bridge is part of the SP's Golden State Route to St. Louis. Here an eastbound piggyback train crosses the Cimarron River. — RICHARD J. COOK

The Pennsylvania Railroad steel truss bridge over the Ohio River (at the falls) at Louisville, Kentucky, is a striking structure which is mounted on stone piers. To obtain greater length and depth at two sections of the bridge, the truss was curved to obtain greater strength at the center, and often called a "Curved Chord Baltimore or Pennsylvania truss." This bridge is now part of the Conrail System. — RICHARD J. COOK

Two great Chesapeake & Ohio System bridges are shown on this page. (ABOVE) Baltimore & Ohio No. 6976 leads a long freight train over the Susquehanna River Bridge at Havre de Grace, Maryland. — H. H. HARWOOD, JR. (RIGHT) The C&O's continuous truss bridge at Cincinnati, crosses the Ohio River. — RICHARD J. COOK

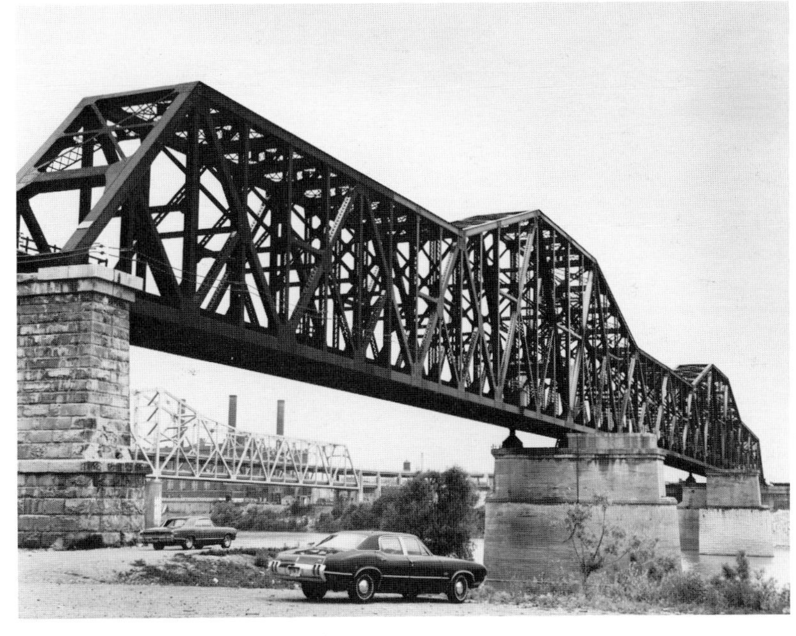

Norfolk Southern (former Norfolk & Western Railway) uses this four-span truss bridge to cross the Ohio River at Kenova, West Virginia. Heavy coal and merchandise trains demand a husky double-track structure. This view is rather deceiving, as the bridge is gigantic. Note that the train crossing the bridge is hardly visible. — RICHARD J. COOK

Southern Railway's (Norfolk Southern) deck truss over Cumberland Lake, near Burnside, Kentucky, was opened to rail traffic on August 3, 1950. This new bridge replaced an older structure now covered by the waters of the reservoir. — RICHARD J. COOK COLLECTION

The Merchants Bridge at St. Louis is operated by the Terminal Railroad Association of St. Louis. The road serves the industrial district on both sides of the Mississippi River and connects and interchanges traffic with 22 railroads. A TRRA diesel takes a long train across the Merchants Bridge. — RICHARD J. COOK

Where large distances must be bridged, a series of simple trusses mounted on a succession of piers solved the problem. As for the continuous truss, the entire truss reacts as a unit, no matter how many supports are placed underneath the span. At the right, the MacArthur Bridge at St. Louis, consisting of three Petit trusses, each 668 feet long, was built in 1918. It consisted of a highway deck and a rail deck underneath. (LOWER-RIGHT) The Kentucky & Indiana Terminal Railroad bridge crossing the Ohio River between New Albany, Indiana, and Louisville consists of a mixture of trusses. (BELOW) The Conrail (former Pennsylvania Railroad) bridge over the Ohio River at Steubenville, Ohio, consists of a 3-span continuous truss. — ALL RICHARD J. COOK

A fast Southern Pacific piggyback freight train clatters across the six-span truss bridge over Amistad Reservoir in Texas, about five miles west of Del Rio, a division point on the railroad. While not spectacular above the water level of the reservoir, it is said that the piers of this structure reach 240 feet to the canyon floor below. — RICHARD J. COOK

New York Central's Alfred H. Smith Memorial Bridge was designed by H. T. Welty and opened to rail traffic on November 24, 1924. The Hudson River is spanned here by two massive subdivided Pratt trusses, one 601 feet long, and the other 409 feet. The structure is 145 feet above the water, ten times higher than the government's limited clearance law. For years the bridge reigned alone over the surrounding countryside, until the New York State Thru Way bridge to New England was built alongside. NYC's (now Conrail's) Selkirk Yard was built at the northern (west) end of the bridge. — RICHARD J. COOK COLLECTION

In 1922 American Bridge Company erected this 700-foot Pennsylvania truss bridge across the Tanana River on the Alaska Railroad. Designed by Ralph Modjeski, the bridge was put together near Pittsburgh, crated and assembled at Nenana near the bridge site. — RICHARD J. COOK COLLECTION

Bessemer & Lake Erie's Allegheny River Bridge is a double-track continuous deck truss built in 1918 to replace a single-track simple span bridge built in 1897. Since the B&LE hauled ore to Pittsburgh and coal to Conneaut on Lake Erie, the new bridge was required to withstand a weight of 8,000 lbs. per square foot. In the above scene, four EMD F-7 units pull an iron ore train toward North Bessemer, Pennsylvania. — BESSEMER & LAKE ERIE (RIGHT) An idea as to the size of the bridge can be gained from the highway which parallels the river. Building the new bridge involved widening the old piers enough to permit erection of the new structure. Once the new bridge was in operation, the old structure was taken down. — RICHARD J. COOK

When steam was still the king of the rails, the B&LE's Allegheny River Bridge saw plenty of action with heavy iron ore trains running south and solid coal trains running to Lake Erie ports. In the view at the left, a Texas type 2-10-4 rolls across the huge Allegheny River Bridge. — BESSEMER & LAKE ERIE

A Texas type 2-10-4 was ample power to handle an iron ore train over the undulating grade of the Bessemer & Lake Erie Railroad. In this view, an ore train rumbles across the Allegheny River Bridge which consists of three continuous spans of 347, 350, and 272 feet. By 1934, when this photograph was taken, the old timber deck had been replaced by a steel and concrete deck with standard ballast. — BESSEMER & LAKE ERIE

The Louisville & Nashville Railroad (now CSX system) built this sturdy five-span truss bridge to carry its north-south traffic over the Ohio River at Henderson, Kentucky. (RIGHT) This great truss bridge carries trains of the Baltimore & Ohio across the Ohio River between Parkersburg, West Virginia, and Belpre, Ohio, on its line to St. Louis. This view looking toward Ohio shows a B&O piggyback train on the bridge. — BOTH RICHARD J. COOK

The McKinley Bridge, built in 1910, was owned by the Illinois Terminal Railroad, an electric interurban line. Like the Merchants Bridge, seen in the distance in the above scene looking north up the Mississippi River at St. Louis, it carried vehicular traffic as well as trains. — RICHARD J. COOK (LEFT) The McKinley Bridge approach from the Illinois side. (BELOW) A view looking west through the bridge itself. The auto highway was on both sides of the truss spans. — BOTH GEORGE KRAMBLES

Looking west on the St. Louis side of the McKinley Bridge. The tracks at the left take off to the downtown passenger terminal. The tracks going straight ahead became Salisbury Avenue and the old route into town via city streets and a few industries along the line. — GEORGE KRAMBLES

A Boston & Maine fast freight train, pulled by four EMD GP-9's, rolls across the Hudson River over a deck truss at Mechanicville, New York. Note the near perfect reflection of the train and the bridge in the quiet waters of the Hudson. —
JIM SHAUGHNESSY

A Duluth, Missabe & Iron Range iron ore train with a Yellowstone type on the head end, crosses a heavy deck truss bridge over the Cloquet River, in northern Minnesota. —
FRANK A. KING

A rare example of a Wadell "A" frame truss bridge was still in use by the Texas & Pacific Railway at Shreveport, Louisiana, in 1977. — RICHARD J. COOK

Nickel Plate's sprightly R-class 4-6-0 No. 155, and its train of two cars, rolls over the Wabash River on a simple two-truss bridge at Lafayette, Indiana. Train No. 22, on its way east from Peoria, Illinois, to Lima, Ohio, was photographed October 1, 1947. — RICHARD J. COOK

A simple through truss bridge on the Denver & Rio Grande Western got narrow-gauge smoke bath this day as engine No. 318 left Ridgeway, Colorado, with a freight bound for Montrose. The scene is on the Montrose-Oura Branch with a connection to the Rio Grande Southern at Ridgeway. Not the bridge was originally built to accommodate standard gauge rails. – JOHN KRAUSE

The Fort Dodge, Des Moines & Southern, an electric interurban line, crossed the Des Moines River on this deck truss bridge just south of Fort Dodge. The road recrossed the Des Moines River once again on a 156-foot high steel viaduct just north of Boone. — WILLIAM C. JANSSEN

Skew truss bridges were common on railroads where the span was entered on a curve in the track. At the left, the Santa Fe Railway's structure over the Los Angeles River was built in 1901. It was washed out by floodwaters in 1935 and then rebuilt. — DONALD DUKE COLLECTION

1656

Pittsburgh's Smithfield Street Bridge, crossing the Monongahela River, was the first and largest structure to employ the Pauli system of lenticular trusses in the Western Hemisphere. In this bridge of two 360-foot through lenticular trusses, Gustav Lindenthal employed steel in the top and bottom cords, piers, posts, diagonal ties and pins. An aluminum floor was added in 1933. Removing some 700 tons of deadweight from the load carried by its trusses. The remainder of the bridge stands as a tribute to a famous designer and the metal which gave Pittsburgh its economy for so many years — steel. The bridge was completed in 1881 and replaced a famous Roebling suspension bridge. It was opened to vehicular traffic in 1882 and the trolley right-of-way was added in 1890. The streetcars abandoned the structure in 1986 in favor of a new rapid transit route across the river. (LEFT) A Pittsburgh Railways PCC car No. 1656 rolls across the bridge toward the Pittsburgh & Lake Erie station stop at the west end of the bridge. (ABOVE) The west gate showing all the decorative fixtures. (RIGHT) The structure and stone piers. — ALL RICHARD J. COOK

The arch bridge is one of the most beautiful of the railroad bridges. An arch bridge support must hold up the weight of the bridge and its load, but this support must also push inward in a horizontal direction to prevent the arch from spreading and flattening out. The five-arch Soo Line bridge over the St. Croix River, near Richmond, Wisconsin, was built in 1911. — RICHARD J. COOK

6

Those Beautiful Arch Bridges

The steel arch bridge came into favor near the beginning of the 20th century and has since been used only on railways where there are deep narrow valleys to be crossed. These bridges are not practical for rail use in other instances because they require extremely massive abutments and foundations in places where there is no natural topographic feature to provide an adequate reaction. In addition, long approaches are often necessary in order to maintain a minimum grade up to the bridge deck level.

But a steel arch bridge was just the thing to use over the deep Niagara gorge when it came time to replace the ancient Roebling suspension bridge.

The Grand Trunk Railway Bridge, solidly braced against the adjoining rock walls was completed in 1897 with Leffert L. Buck as chief engineer and Richard S. Buck as resident engineer. Known as the Niagara Falls Railway Arch Bridge, it was constructed by the Pennsylvania Steel Co. of Steelton, Pennsylvania, and utilized a 550-foot span arch, two-hinged, spandrel-braced. It carries the double-track railway on its top deck and highway traffic on the lower deck just on top of the arch itself.

This type of bridge proved to be very popular in the Niagara gorge because of the topography, so that in 1924, when the Michigan Central decided it needed a bridge just a few hundred feet upstream from the Grand Trunk's arch, to handle traffic over the gorge, it replaced its cantilever span with a 640-foot long steel arch. It, too, carries two tracks on the deck at 13-foot centers, and was designed for erection by the cantilever method. It belongs to the spandrel braced type of steel arch with the lower chord conforming to a parabolic curve and the top chord horizontal. The base of the rail is 240 feet above low water level. The bridge is now operated by and is the property of the Conrail System.

In the early part of the century, the Soo Line replaced an old bridge with a five-arch bridge over the St. Croix River near New Richmond,

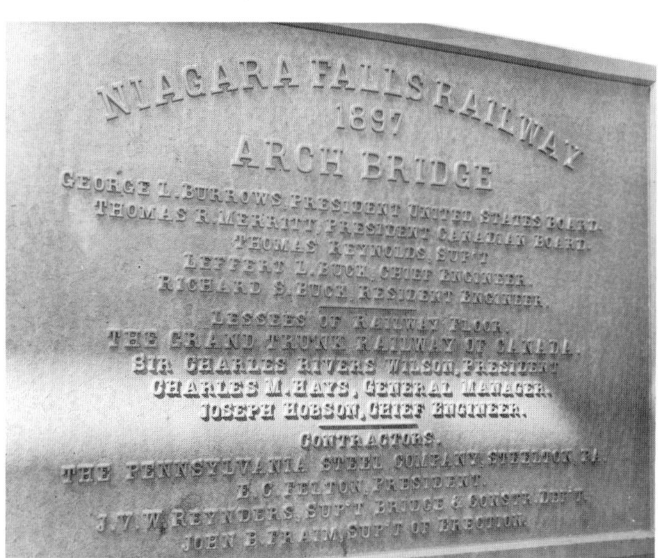

Two Niagara Gorge bridges are shown in this photograph. The first is the Grand Trunk arch bridge, and the one directly behind it is the old Michigan Central cantilever, since replaced with an arch bridge. The old electric *Niagara Gorge Route* railway track may be seen at the left of the scene. At one time the tourist could reach a vantage point to see the falls by taking a thrilling trolley ride to just a few feet from the broiling currents of the tumultuous waters of the gorge. (LEFT) The bronze plaque belongs to the Grand Trunk Railway bridge, built in 1897 to replace the famous Roebling suspension bridge. — RICHARD J. COOK

The Niagara Falls Railway arch bridge, built by the Grand Trunk Railway in 1897, is a 550-foot span carrying two tracks on the upper deck and vehicular traffic on the lower deck. Leffert L. Buck was the chief engineer on the project and Richard S. Buck was the resident engineer. Rail traffic between the Canadian National and Conrail is carried on daily as freight trains run back and forth to the respective railroad yards at Niagara Falls on the Canadian side and Buffalo on the American side. Customs inspection is made on both sides of the border, and it is necessary to re-classify on both sides due to the heavy freight traffic running between American rail points and Hamilton and Toronto, Ontario, on the Canadian side. — BOTH RICHARD J. COOK

In 1924 the New York Central (Michigan Central) replaced its two-light-for-the-modern-traffic cantilever bridge with a 640-foot long steel arch structure, complementing the Canadian National (Grand Trunk) arch bridge just a few hundred feet downstream. The base of the rail of this bridge is 240 feet above the low water level. — RICHARD J. COOK

Canadian Pacific's *Atlantic Limited* now operated by Via Rail, crosses a combination truss and arch bridge over the Reversing Falls of the St. John River at St. John, New Brunswick. — RICHARD J. COOK

Wisconsin, in 1911. The old alignment with sharp curves and steep grades was eliminated by the new main line bridge across a wide valley. The overall length of the bridge is 2,682 feet, with approaches of 340 feet to the east, 560 feet to the west. The arches measure 358 feet each. This picturesque bridge is singularly graceful in appearance.

Four years before the St. Croix bridge was completed, the engineers of the Pennsylvania and New Haven railroads had prepared preliminary plans for what was to become the greatest of all American arch spans, both in total size and in magnificent appearance. The Hell Gate Bridge at the upper end of the East River in New York City was originally conceived as a part of the Pennsylvania Railroad terminal project of 1903-10. The New York Connecting Railroad was organized in 1911, construction of the new line began in 1912 and the first train moved over the new link between those two railroads in April 1917, making for a smooth connection between New York City, Boston, and Washington. Ten years later the route was electrified. The huge bridge — the longest arch in the world at the time of its construction — brought international attention.

Ground was broken on March 12, 1912, for the mighty Hell Gate Bridge crossing New York's East River. The work to complete the structure took four and one-half years. During April 1917 the bridge was opened to traffic and the first trains were operated by steam locomotives. It was ten years later before the route over Hell Gate was electrified. Upon completion this bridge became the longest arch bridge in the world. — LIBRARY OF CONGRESS (BELOW) Built like a fortress, the husky end of the arch looks like it should include a drawbridge and a moat. — RICHARD J. COOK COLLECTION

The Liberian registered freighter *Rhenonia,* based at Monrovia, is about to pass under the famous Hell Gate Bridge as a GG-1, with Penn-Central initials on its side, whisks a short passenger train through the arch en route to Penn Station. GG-1 electrics were often run over the bridge and operated as far north as New Haven, Connecticut. — RICHARD J. COOK

Hell Gate is a skyscraper of a bridge with its huge masonry portal towers built upon bedrock 70 feet below the ground and soaring up above the treetops. The towers dominate the entire East River skyline. — RICHARD J. COOK

The Hell Gate Bridge, the triumph of Gustav Lindenthal, is a fitting monument to its designer. The bridge, erected in two years, 1914-16, was a remarkable feat of construction made possible by the simplicity of the arch structure and the relatively small number of its separate pieces. the overall span of the arch is 1,017 feet between abutments at the deck level, while the clear span from center to center of the skewback hinges is 977 feet.

The huge four-track structure dominates the north East River skyline and is notable for its huge masonry portal towers built upon bedrock about 70 feet below mean water level. Trains on its deck look like toys and they ascend and descend long approaches on either end. To travel across the bridge is to feel that you have lost contact with the earth below and that you are floating through space with a marvelous view of the Manhattan skyline to the west and an aerial view of the Long Island shoreline to the east. Actually, you are about 135 feet above high water. The total cost of the longest railroad arch bridge in North America — if not the world — was $28 million.

The New Haven Railroad, which also operated its trains over the Hell Gate Bridge, was electrified in 1927. In this photograph, No. 0308, one of the original electric motors, hauls its train across the four-track bridge. — DONALD DUKE COLLECTION

A 1958 "Royal Train" with Princess Margaret aboard, crosses the Canadian Pacific's Stoney Creek arch bridge in British Columbia. — NICHOLAS MORANT/ CANADIAN PACIFIC

To be introduced to another notable steel arch bridge, we must go to Canada, to the Canadian Rockies to be specific — the main line of the Canadian Pacific Railway. The Stoney Creek Arch replaced a timber trestle installed when the Canadian Pacific first opened in 1885. At the time it was the highest wooden bridge in the world.

The Stoney Creek Arch is located some 3,000 feet above sea level on the eastern slope of the Selkirk mountain range immediately east of the summit and the famous Connaught Tunnel. The second Stoney Creek bridge consisted of two single steel arches, one on each side of the line, with a span of 336 feet and a rise of 80 feet in the form of a segment of a circle, pin connected at the springline and crown, with flanking deck truss spans to the abutments. The bridge was 300 feet above the gorge bed, making it among the highest bridges. In 1928 reinforcement was required in order to accommodate increased loading. This was done by building similar arches outside of, and parallel to, the existing ones, replacing the deck trusses by plate girders supported on new steel bents between the old and new parallel arches at each end of their panel points, by replacing the flanking spans with deck plate girders supported on new bents carried to rock, and by the existing abutments. The alteration, done in the field in 1929, took less than five months and without interruption to traffic.

Three other steel arch bridges of note are the Santa Fe's Canyon Diablo bridge in Arizona, the Crooked River Arch in Oregon and the Hurricane Gulch bridge on the Alaska Railroad.

Oregon Trunk's arch bridge over the Crooked River at Terrebonne, Oregon, was one of the highest bridges in the world when constructed in 1912. Spanning a canyon 350 feet wide and 320 feet deep was accomplished by two hinge-steel arch spans erected by cantilevering out. While the bridge was under construction, a good deal of the supplies were lifted from the valley floor by a traveller crane. Erection was under the supervision of Ralph Modjeski. The Oregon Trunk became part of the Spokane, Portland & Seattle Railway, and now forms a part of the Burlington Northern. — RICHARD J. COOK COLLECTION

Another arch bridge of significance is the 918-foot structure on the Alaska Railroad spanning Hurricane Gulch. Located at milepost 284.2, the arch is 296 feet above the valley floor. — RICHARD J. COOK

The Crooked River Arch is located on the Oregon Trunk Railway at the crossing of the Crooked River canyon about 26 miles north of Bend, Oregon. Here, the single-track span, 340 feet in length, crosses the waters of the Crooked River 320 feet below (two feet less than the Pecos River bridge). Erected under the supervision of Ralph Modjeski, who at that time was chief engineer of bridges for the Oregon Trunk, the bridge was completed in 1912. The site, with its sheer cliffs, naturally adapted itself to an arch span as the opening was comparatively narrow, the walls being capable of taking unlimited thrust when excavated below the cracked and weathered surface rock. The depth of the excavation was so great that cantilever erection was necessary.

The third bridge, on the Alaska Railroad, is located at Mile 284.2 and spans Hurricane Gulch. The bridge is 918 feet long and measures 296 feet from the base of the rail to the creek level below.

Canyon Diablo, 26 miles west of Winslow, Arizona, pierces the flat plateau of Arizona to a depth of 250 feet, with sheer yellow and white limestone rock walls rising almost perpendicularly from the narrow river bed. Here four Santa Fe blue and yellow diesels roar an eastbound fast freight across the 300-foot two-hinged arch bridge. The structure is flanked by a 120-foot double-track simple deck truss. This bridge, the highest structure on the Santa Fe System, was opened to traffic in 1947. — DONALD DUKE

Canyon Diablo was first crossed by the Atlantic & Pacific Railroad in 1882. The bridge iron, which was prefabricated in New York, was shipped west in sections. The structure was designed to hold 30 times the weight of the trains that crossed it. Obviously locomotives and cars were not all that heavy in 1882. The iron viaduct bridge, completed in July 1882, soared 222½ feet above the canyon floor. Supporting the 560-foot long structure were iron bents and towers. In this scene a westbound passenger train crosses the bridge, which was a "must" stop for all tourists going west. — DONALD DUKE COLLECTION

Canyon Diablo is a gash 225-feet deep in the arid tableland of northern Arizona. The bridge replaced two earlier bridges, each of which was of a viaduct type with deck trusses or girders spanning between high steel towers. Built between 1946 and 1947, the double-track arch structure was opened for high-speed main line service in September 1947. It is said to be capable of carrying the heaviest equipment at high speeds.

The Canyon Diablo bridge incorporates a 300-foot two-hinged arch flanked by a 120-foot simple deck-truss span at each end. Each half of the arch was erected by cantilevering it out from the arch piers to join in the center.

Steel arches, cantilevers, continuous and simple trusses have all served railroads well in the 20th century. The truss form has become pretty much the basic style for the larger railroad bridge and these bridges have demonstrated well their utility and aesthetic character; they have been a proven success.

The eastbound *California Limited* pulled by locomotive No. 1933 crosses the second Canyon Diablo bridge in the early 1930's. Completed in 1900, this 550-foot gauntlet track structure was strengthened twice before the arch bridge was completed. — DONALD DUKE COLLECTION

A bridge that is hoisted while remaining in a horizontal position is called a vertical lift bridge. This type of structure has become popular because of its ease of operation and reliability. The longest vertical lift bridge is this one over Arthur Kill at the south end of Newark Bay, New Jersey, connecting Staten Island by rail. Completed in 1959, this Chessie System bridge is 550 feet long and replaced a swing bridge operated by Baltimore & Ohio subsidiary Staten Island Rapid Transit. — RICHARD J. COOK

7

Bridges That Move

Bridges that move — that can open in a few seconds, clear a channel for passing watercraft and then close to form again part of the roadway — became almost a railroad specialty in their first days. Since the late 1800's such bridges evolved into several different types or styles. All had one thing in common: they were the most economical type to build, demanded little change in railroad grade and offered a channel adequate for the passage of most vessels.

Four types of movable spans have been used with success by railroads:

1) The *swing* bridge is one in which the swing or draw span pivots on a pier in the center of the river. Usually the channel passage is at one side or the other of the "draw."

2) The modern *bascule* bridge had its origin in the medieval drawbridge, used to provide passage over the defensive moats that encircled castles. Sketches for counterweighted bascule bridges were found among notes made by Leonardo da Vinci as early as 1500. The *single-leaf bascule* bridge, or "jackknife," offers a vertically clear channel where necessary and is an improvement over the midstream pier swing bridge where the channel is not too wide. The swing span, however, was said to be less expensive.

3) A very rare bridge for railroad use is the *double bascule*. The best example of this type is the Canadian Pacific bridge over the United States' ship canal between Lake Superior and Lake Huron on the St. Mary's River connecting the two Sault Ste. Maries — Ontario and Michigan.

4) The *vertical lift* bridge is the structure now in favor for rail use where a movable span is called for. The reasons are these: (a) simple to design, (b) reliability, (c) ease of operation, (d) short operating time, (e) economy of power, which makes it more economical than a swing bridge, (f) a long span may be more economically constructed than either a long swing or a long bascule bridge, (g) affords a much greater channel width than a swing, (h) no center pier — as

After this Burlington Northern passenger train cleared and discharged its passengers at Vancouver, Washington, the Columbia River swing bridge, shown in the background, opened up once again for river traffic. — RICHARD J. COOK

with a swing span — to interfere with navigation, (i) less chance than either a swing or bascule for collision with boats, (j) possibility of interchanging spans. A bridge of several spans, having one arranged as a vertical lift, may be constructed so as to permit the moving of the towers and lifting machinery from one span to another whenever this is advisable due to a shifting channel. This also holds true when a channel has to be widened.

The vertical lift bridge has had its chief development in the United States. It has rendered the older movable bridge types practically obsolete. The vertical lift and the single-leaf bascule are competitive types, although for spans over 350 feet, the vertical lift appears to have the field to itself.

The world's longest and heaviest swing bridge is owned by the Santa Fe Railway, over the Mississippi River at Fort Madison, Iowa. It is a combination rail/highway bridge linking Fort Madison with Niota, Illinois. The Santa Fe's Illinois Division main line runs at deck level while the highway utilizes the upper portion of the bridge. The main river crossing consists, from east to west, of four fixed through truss spans of 270½ feet and a draw span with two equal arms of 265 feet 10 inches, the distance being measured center to center of piers. All piers are set square with the center line of the bridge with the exception of the two rest piers for the draw span which conform to the skew of the channel. The east approach consists of nine spans of girders 100 feet, center to center of piers and the west approach of nine spans of girders 80 feet and one span 102 feet, center to center of piers.

Construction on the bridge began in 1926 and was completed in 1927. It replaced a bridge built in 1887 that was damaged by fire in 1923. The draw span of the present bridge measures 531 feet in length. The draw was first opened at 3:40 P.M. July 28, 1927 for the Navy's subchaser SC-64 from the Burlington, Iowa Naval Station. The first train crossed the bridge on July 25, 1927.

The toll bridge for the highway traffic is operated around the clock by the bridge operator who is housed in a booth tucked inside the pivoting swing span section. U.S. Coast Guard requirements state that the draw span shall be opened promptly when a signal is received from an approaching vessel which cannot pass under the closed draw, except when an approaching train is so close that it cannot be safely stopped before reaching the bridge. Aside from his toll-collecting duties, the operator is also in charge of the swing span, operating it and maintaining radio contact with rail and marine vessels to coordinate movements. He usually knows 20 to 30 minutes in advance of a span opening. Approaching highway traffic is warned by a flashing red light atop the bridge and the sounding of a klaxon or siren.

The first railroad bridge over the Mississippi River was built at Davenport, Iowa, in April 21, 1856. It was replaced by the Davenport, Rock Island & Northwestern Railroad in 1900 with a seven-span truss bridge, plus one 442-ft. swing span, making it one of the longest bridges of its type at that time. Since then there have been many swing bridges put to use for rail traffic.

Since 1919 the record for the single-leaf bascule spans has been the 260-foot span of the St. Charles Air Line Railway bridge at 16th Street, Chicago.

Apparently the only double-leaf bascule bridge for railway operation in North America has been the Canadian Pacific's bridge at Sault Ste. Marie, Ontario. This 336-foot span is unique in that it acts as a simple truss when closed, whereas double-leaf bascules are generally designed to act as two cantilevers locked together at the center. Each leaf or section of the span consists of a 168-foot arm and a 45-foot tower. It is accurately balanced by counter-weights containing about 550 cubic yards of concrete each and weighing over 1,000 tons. Both leaves, or halves, can be fully opened for large, deep draft vessels using the center of the canal. On the other hand, for the passage of small vessels, only one leaf can be opened. The Sault Ste. Marie bridge went into service in September 1914, and was an important lifeline between Canada and the United States during World War I when iron ore was transported by rail for the war effort.

The bridge was severely damaged in 1941 when a locomotive travelling from north to south ran into the canal when the north half of the bridge was only partially opened. In the ensuing rebuilding, electrically-operated shear pins were installed to ensure that the ends of the two leaves remained securely held in place when the bridge was in the closed position.

The longest vertical lift bridge, ironically, is used by a lightly travelled industrial spur line of the Baltimore & Ohio over Arthur Kill at the south end of Newark Bay in New Jersey. Completed in 1959, this 550 foot, (center to center of its bearings) bridge replaced a 70-year old swing span and became the longest vertical lift in the world. The bridge was operated by former B&O subsidiary Staten Island Rapid Transit Railroad Company.

The next longest vertical lift in North America is the Buzzard's Bay Bridge over the Cape Cod Canal in Massachusetts. Built in 1935, its span measures 544 feet in length.

Next comes a 542-foot double-track bridge over the Delaware River at Camden, New Jersey. This is follwed by the 408-foot lift span of the Missouri-Kansas-Texas at Booneville, Missouri, with service beginning on February 1, 1932. The bridge, with a 30-foot lift over the Missouri River, replaced a swing bridge, originally built in 1874 and has since been upgraded. It has survived many treacherous Missouri River floodings and proven itself the utilitarian heavyweight it was designed to be.

Swing Bridges

A bridge that pivots to align itself with the water traffic and presents a minimum width across the water traffic path is called a swing bridge. The world's longest and heaviest swing bridge is owned and operated by the Santa Fe Railway, over the Mississippi River, at Fort Madison, Iowa. (LEFT) This photograph shows the two-deck arrangement of this bridge built to handle not only the heavy rail traffic of the Santa Fe, but vehicular traffic on the upper deck as well. (RIGHT) A westbound Santa Fe freight has just crossed the Fort Madison bridge and heads into the division point at Fort Madison. — BOTH RICHARD J. COOK

Even the busy Northeast Corridor rail line has swing bridges. This unusual bridge swings at Cos Cob, Connecticut, on the old New Haven Railroad line between New York and Boston. In addition to the many commuter trains which use the bridge, as seen in this view, Amtrak passenger trains between Boston, New York, and Washington, D.C., pass this spot at regular intervals. From the position of this photograph it is hard to see how this bridge can swing, but it does. Note that the overhead is not carried over on the swing bridge itself. — H. H. HARWOOD, JR.

Hannibal, Missouri, is the hometown of Mark Twain (Samuel Clemens) who wrote the Tom Sawyer and Huck Finn books in addition to tales about life on the Mississippi River. The town is also home to a most interesting swing bridge built by the Wabash Railway to accommodate river barge traffic. (ABOVE) The bridge may be seen with its six-spans and swing section. (LEFT) Not only is the bridge unique, but on the Hannibal side of the river, the track also plunges into a short tunnel under the high cliffs before curving into Hannibal itself. (BELOW) Looking south, or down river to the Hannibal swing bridge which is at the right in this photograph. The bridge and its railroad are now part of the Norfolk Southern System. — ALL RICHARD J. COOK

The aerial view of the Chicago, Rock Island & Pacific Railroad swing bridge at Little Rock, Arkansas, illustrates the operation of a swing bridge. The operator, nestled in a little house high above the center of the swing span, has full control of the bridge. — RICHARD J. COOK COLLECTION (LEFT) Clinton, Iowa, the heavy-duty swing span of the Chicago & North Western. Here the opertor runs the swing span from a similar in-truss control house. — RICHARD J. COOK

The Rock Island six-span swing bridge at Davenport, Iowa. Note the railroad line on the upper deck and an auto highway on the lower deck. — RICHARD J. COOK

In order for a swing bridge to swing freely, there has to be a gap in the rail connection at both ends of the swing portion of the bridge. The break is nearly a foot wide and in order for a train to pass smoothly when the bridge is locked closed, a blade of rail is moved back and forth by compressed air. The blade is similar to that of a switch. The movement of the blades is controlled by the bridge tender or operator. (LEFT) The rail connection in the closed position showing the blade. (RIGHT) A view showing the gap and the rail connection in the closed or locked position. The compressed air operated crank may be seen at the top of this picture. — BOTH RICHARD J. COOK

The swing connection of the Milwaukee Road swing bridge at Sabula, Iowa. Note the home signal mounted on a bridge support, however, it is not a part of the swing portion itself. When the bridge is about to swing open to allow for river traffic to pass, the operator sets the signals some miles distant to amber, with a red signal some hundred feet from the bridge itself. Should these amber signals be ignored, this second red signal will warn the engineer he had better stop or take a bath in the river. — RICHARD J. COOK

The Oregon-Washington Railroad & Navigation Company's (Union Pacific) lift bridge over the Willamette River at Portland, Oregon, has two decks, with the railroad on the lower deck. This rail deck can be raised separately for some distance in order to permit the passage of small vessels, tugs and lighters. The whole lifting span, including the rail deck, can be raised to permit the passage of ocean-going ships. In the above scene, Union Pacific's streamliner train No. 109 *The City of Portland,* has left Portland Union Terminal on October 19, 1967, crosses the Willamette River Lift Bridge and heads for Chicago. The Southern Pacific also uses this bridge to reach the Portland Union Terminal. — HENRY R. GIFFITHS

One of the longer vertical lift bridges used only for railroad use is the 2,633-foot long Harry S. Truman bridge over the Missouri River at Kansas City. The bridge was operated as a joint effort by the Milwaukee Road and the Rock Island. The current operator of the bridge is the Soo Line.—RICHARD J. COOK

Vertical lift bridges are also ideal for crossing canals. Canadian National's Welland Canal Bridge is usually up most of the time to accommodate the heavy ship traffic that passes through on its way between Lake Erie and Lake Ontario. Many smaller foreign vessels use this route from the St. Lawrence Seaway to the Great Lakes ports on lakes Erie, Michigan, Huron, and Superior. The lift bridge is lowered for local rail traffic. — RICHARD J. COOK

The Buzzards Bay Lift Bridge over the Cape Cod Canal in eastern Massachusetts, is the second longest vertical lift bridge in the United States. The 544-foot span was erected in 1935. In the open position the bridge provides a clearance of 139 feet above mean sea level and a clear width of 500 feet between fenders. The towers are so designed that the lift span is maintained in the open position except for infrequent rail movements. The bridge was built by the New York, New Haven & Hartford, and is now part of the Conrail system. — RICHARD J. COOK

Under a threatening sky, the Northern Pacific's *North Coast Limited* rolls across the Columbia River Lift Bridge at Pasco, Washington. — RICHARD J. COOK

The Norfolk & Portsmouth Belt Line's lift bridge over the southern branch of the Elizabeth River at Norfolk (Virginia) Harbor is 1,153.5 feet long, including the approaches. The lift portion of the span, a 384-foot long Warren-type truss, provides a vertical clearance of 145 feet when fully raised. The channel opening is 300 feet wide. The bridge was built in 1958 by American Bridge Company to replace a 277-foot swing span which had been erected in 1898. (LEFT) The bridge in the closed position. (RIGHT) The lift span has opened part way to allow for the passage of a barge. — BOTH RICHARD J. COOK

When it was built in 1932, this Missouri-Kansas-Texas Railroad lift bridge was said to be the longest (408 feet) vertical lift span for railroad use. The bridge crosses the Missouri River at Boonville, Missouri. The lift span, when opened, gives a headroom of 57 feet at low water. In addition, there are three 300-foot fixed spans, one of 247 feet and a 60-foot through-girder span. This photograph was taken when the river was at near flood level. (INSET PHOTOGRAPH) An up-river view showing a side view of the MKT bridge. — BOTH RICHARD J. COOK

The former Pennsylvania Railroad lift bridges at Newark, New Jersey, see plenty of railroad action. (LEFT) A Port Authority Trans-Hudson (PATH) train arrives at Newark from lower Manhattan. (BELOW) A Pennsylvania Railroad GG-1 electric locomotive, operated by Penn-Central when this photograph was taken, crosses through the Newark lift bridges and then pulls to a stop at Newark Station. — BOTH RICHARD J. COOK

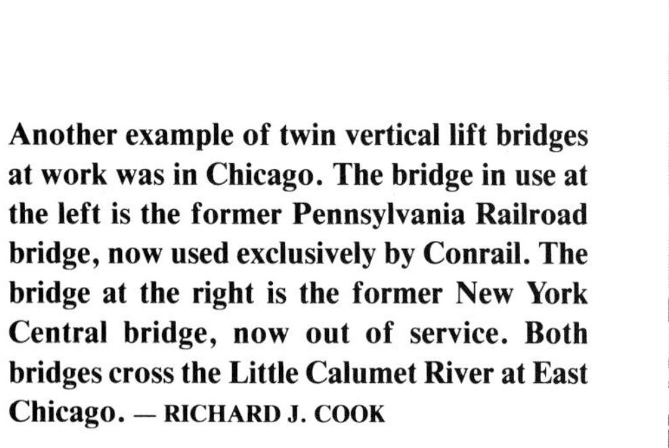

Another example of twin vertical lift bridges at work was in Chicago. The bridge in use at the left is the former Pennsylvania Railroad bridge, now used exclusively by Conrail. The bridge at the right is the former New York Central bridge, now out of service. Both bridges cross the Little Calumet River at East Chicago. — RICHARD J. COOK

Southern Pacific's bridge over Carquinez Straits of Suisun Bay, 31 miles northeast of San Francisco, was an engineering triumph for its day. It eliminated the use of two huge train ferries. The 5,603-foot bridge has ten main bridge piers and 22 piers on the viaduct approaches. The approaches required six miles of new railroad, including the spectacular steel bridge with its 328-foot lift span, and a riveted Warren truss span weighing 1,580 tons. This gives a 135-foot clearance above high water. (ABOVE) Daylight type 4-8-4 No. 4462 rolls the Oakland-bound *Cascade* over the north end of the bridge. — SOUTHERN PACIFIC (LEFT) FT diesel unit No. 6460 heads a long freight over the bridge. BART GREGG (LOWER LEFT) At the dedication ceremony of the bridge on November 1, 1930, old *C.P. Huntington* was brought out of retirement to be the first locomotive over the bridge. — SOUTHERN PACIFIC (BELOW) Map showing location of the bridge, the last simple truss bridge to be built in the United States at a new site.

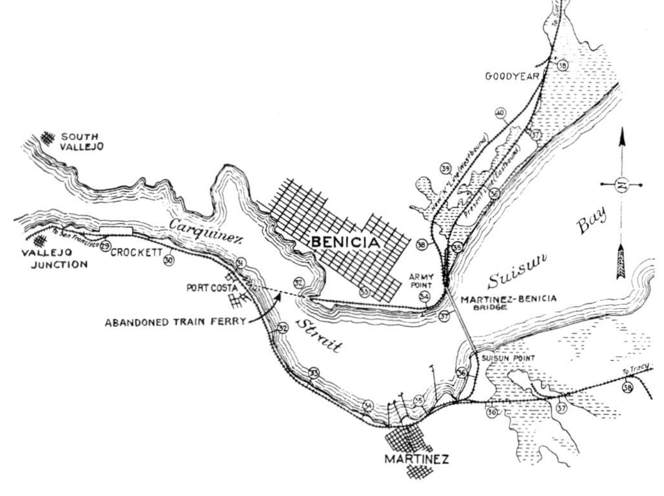

Under a heavy sky, Amtrak's *Silver Star*, a New York-Miami streamliner, rolls across Lake Monroe and passes through a very modern lift bridge on the former Atlantic Coast Line route. — RICHARD J. COOK

Bascule Bridges

A Baltimore & Ohio transfer train crosses the Cuyahoga River at Cleveland, Ohio. This bridge is a perfect example of a single-leaf bascule bridge. — RICHARD J. COOK (LEFT) This old New York Central bascule bridge, on the pier track at Sandusky, Ohio, was hand operated by the crew of the train switching in that area. The controls were located in the shanty at the right. — H. H. HARWOOD, JR.

The 260-foot St. Charles Air Line single-leaf bascule bridge at Chicago sees Amtrak trains (among others) operating over it. All trains from Chicago's Union Station to the Illinois Central Gulf tracks make use of this bridge. It is one of the longer bascule bridges in the country. — RICHARD J. COOK

It is unusual to find an electric interurban drawbridge that lifts from both sides. In this scene on the Illinois Terminal Railroad, car No. 282 rolls across this double bascule drawbridge over the Illinois River at Peoria. — PAUL H. STRINGHAM

What may be the only double bascule bridge still in use for rail service in Canada is this Canadian Pacific bridge over the ship canal at Sault Ste. Marie, Ontario. Here lake boats and foreign ships pass from Lake Huron to Lake Superior. — RICHARD J. COOK COLLECTION

A Los Angeles bound Pacific Electric interurban train rolls through the bascule bridge on the San Pedro to Los Angeles line. The bridge provided access to the Inner Harbor of Los Angeles Harbor at San Pedro. — DONALD DUKE

Great Northern Railway's local bound for Spokane, thunders across the Salmon Bay drawbridge over the Lake Washington Ship Canal. Lake Washington is just east of the city of Seattle. — DONALD DUKE COLLECTION

These three rolling bascule bridges at Chicago were built by the Scherzer Rolling Lift Bridge Co. Although they no longer lift for traffic on the Chicago Drainage Canal, they see plenty of rail action as many roads send their switching and transfer trains through the bridge. — RICHARD J. COOK

Can you name a railroad line crossing the Sault Ste. Marie Canal that has a double bascule bridge and a lift bridge built back to back? If you guessed the Canadian Pacific at Sault Ste. Marie, Ontario, you were right. — RICHARD J. COOK

ver wonder what the view would be from the con- ol house of a bascule bridge? Well, on the oppo- te page, you can get a good idea of the view from ne of these crow's nests. About all the operator n see is the roofs of cars and locomotives, plus all e braces of the bridge. — LIBRARY OF CONGRESS RIGHT) The Baltimore & Ohio (Chessie System) ngle-leaf bascule bridge over the Calumet River at hicago, resembles some monster in a science fic- on film. — RICHARD J. COOK

Anyone with acrophobia would certainly not wish to walk across the gigantic Lethbridge viaduct of the Canadian Pacific over the Oldman River at Lethbridge, Alberta, Canada. The structure is a modest 5,328 feet long and stands 313 feet high. — RICHARD J. COOK

8

Bridges on Stilts Trestles and Girders

Those spidery-looking structures that we often see carrying railroads across rivers or canyons are the modern version of the old wooden trestle, bridges built with a latticework of wooden beams all bracing and supporting the track above it. The old trestle was easy and relatively fast to put up — and it was cheap. But because of its slender construction and rather light weight, it could not be used to carry heavy trains without causing a severe speed restriction. Many, however, are still in use. The longer trestles are rugged but still call for slow speeds for their rail traffic.

The steel viaduct consists of longitudinal beams supported by a series of great steel towers. Although all metal viaducts are basically the same, the combination of great height and the repeating pattern of steelwork can give them a breathtaking quality and an air of grace that cannot be found in the heavier truss design.

The greatest of all these steel viaducts is without a doubt the Lethbridge Viaduct at Lethbridge, Alberta, built by the Canadian Pacific Railway.

No one has been able to come up with an imposing name that seems suitable for Canadian Pacific's impressive viaduct. Initially, some people called this high railway bridge the Big Bridge over the Belly River but, nevertheless, that didn't seem to catch on. In those days, too, many Victorian Lethbridge folk found the word "Belly," an Indian name, coarse and changed the river's name to "Oldman," another Indian name.

In the early years the designers called it the CPR Viaduct or the Lethbridge Viaduct, while other sources named it the CPR Trestle. In 1976, CP Rail identified the bridge as Viaduct 1.1, Crowsnest Subdivision.

Whatever its name, "Wonder of the World," completed in 1909, is still the highest and longest structure of its kind, with a length of 5,328.6 feet and a maximum height of 313.7 feet. It accommodates today's heavy trains with ease.

Before its constuction, westbound trains

An eastbound Canadian Pacific freight rattles across the Lethbridge viaduct and then pulls into the yard at Lethbridge. The structure, according to the Canadian Pacific, is the highest and longest steel bridge of its kind in the world. Spectacular to say the least. — RICHARD J. COOK

The Lethbridge viaduct was built by the traveller method: a crane moving out over the rails laying girders across the steel piers as it goes. Here are two views taken of the construction of the bridge in 1909. Note the two men jumping off the last girder in the above view. — BOTH RICHARD J. COOK COLLECTION

from Lethbridge ran a circuitous route along the Oldman River to a point about five miles upstream where a crossing was made. From there, the rail line climbed a steep grade, crossed 24 bridges and navigated curves equal to five circles. The new route via the new bridge shortened the trip by more than five miles, reduced the number of bridges to two and lopped one hour off the schedule time.

The viaduct consists of 44 plate girder spans 67 feet long, 22 plate girder spans 98 feet long, and a riveted deck lattice truss span 167 feet long. It is carried on 33 riveted steel towers, rigidly braced. The substructure consists of concrete piles supporting concrete pedestals. The sheet steel piles under the land piers vary in depth from 12 to 20 feet.

Column footings had to be adequately anchored to piers because of the uplift from high winds. Anchor bolts 2½ inches in diameter and nearly nine feet long were used in such a way that one corner of a tower was fixed, while the other three were free to move with expansion or contraction of the steel. Allowance for movement was provided by a system of sliding plates, permanently lubricated with graphite.

Because of the bridge's unusual height and the frequency of severe winds, through spans were used instead of the deck plate girder type. The track was nestled between two eight-foot high girder spans instead of running on top of them — making it impossible for derailed cars to fall over.

These same high winds also played havoc with preliminary work and measuring operations. An unsupported steel measuring tape couldn't be used, so a new device was made on the spot from a well-seasoned 16-foot long cedar two-by-four with brass plates attached. This resulted in a highly accurate 15-foot measuring rod. Even the plumb bobs had to be protected by wind shields.

To build the mammoth steel structure, a special erection traveller had to be made at a cost of $100,000 — a lot of money by 1908 standards. It was made entirely of steel, except for flooring and engine house. The whole thing stood 60 feet high, took a month to build and weighed 712,000 pounds. Eleven miles of cable were used for lowering and raising the column material.

For safety, men worked from two assembly cages which were put into place by 85-foot booms and temporarily bolted to the steel columns. To expedite the riveting jobs, a smaller riveting traveller was also used, from which cages were suspended from an overhang.

When the job was completed, the structure was given two coats of paint — 7,600 gallons of it. The bridge was so well constructed that it was 49 years before any major repairs were needed.

Union Pacific's gigantic Jose Bridge, on the Spokane Line, crosses the Snake River in this 1966 photograph, as train No. 19 rumbles over the 3,920-foot structure. The piers of the bridge are being raised because of a dam under construction downriver. Cofferdams protect the piers while the job is in process. — HENRY R. GRIFFITHS

Wooden trestles — large ones such as this — are fast disappearing. Here Camas Prairie train No. 344, handled by Union Pacific No. 2881 a 4-6-2, crosses over U.S. Highway No. 395 near Craigmont, Idaho, in 1947. The Camas Prairie was a joint Union Pacific/Northern Pacific operation which used its owners' cars and locomotives. The classic coach, with a Northern Pacific letterboard, is a beautiful wooden car. — HENRY R. GRIFFITHS

Wood Trestles

Webster's *New International Dictionary* states that the word "trestle" means a braced framework of timbers, piles, or steelwork, of considerable height, for the carrying of a railroad over a depression. Timber work is what comes to mind when a railroad historian thinks about a trestle, while a steel trestle is considered a viaduct. In any case, this chapter does cover both of these trestles — wood and steel.

Wood trestles were particularly common in North America for many years since timber reserves had not as yet been depleted. They were also attractive to early railroad builders because they used locally available timber, and could be assembled fairly quickly at a moderate labor cost. During the construction era it took months to build a large earth fill using horsedrawn scoops or mine-size dump cars. With wood trestles the railroad could be completed quickly and if it was necessary, the trestle could be filled in to form an earth fill. In this century, steel, concrete, and especially prestressed reinforced concrete not only saved maintenance costs, as compared to timber, but was simpler to erect.

Timber trestles were built in two types known as pile trestles and frame trestles, and were made in sections, called bents, which were spaced 12 or 14 feet apart.

Pile trestles were used largely in stream beds and swampy spots where good foundations for framed trestles could not be secured. On most railroads one pile was placed in the center of the roadbed and the outer piles placed from 24 to 48 inches on either side of it. On medium height trestles four piles were used, the two inner ones being spaced three feet apart, center to center, and the outer piles 26 inches, center to center, on either side of the middle ones.

The piles were driven into the ground by a steam driven pile driver to bed rock, or until a secure or solid bottom was hit. The piles were then sawed off to the required height above ground. The pile driver generally had a crew of eight and could drive and cut from 25 to 36 piles per day during a ten hour period. A cap was then bolted to the top of the pile. Pile trestles less than nine feet high were seldom braced, but where the height exceeded this they were braced on each side with 3 × 6-inch scantlings placed diagonally across both sides of each row of piles.

Framed trestles were both round or squared timbers. The frames, or bents, consisted of four supports, or legs, made of timbers from 15 to 18 inches in diameter or 10 × 12-inch squared timbers. The legs rested on a timber called a sill to which they were drift-bolted. Sills varied in length according to the height of the trestle. In many cases the trestles were bolted to concrete piers or mud sills to provide a firm support on the ground. The tops of the legs were covered with a cap 12 or 14 feet on which the stringers rested.

Timber trestles are still widely used for small bridges, however, they are heavily creosoted to prevent rot and termite destruction.

During the turbulent era of railroad logging on California's Pickering Lumber Company operations, *Pacific Coast* type Shay No. 11 emits a cloud of oil smoke as it puts power to the rails. The climb up Schoettgen Pass was riddled with trestles just like this wood structure. — AL ROSE

Colorado's famous Rio Grande Southern narrow-gauge line had many spectacular wood trestles. (LEFT) The long trestle at Ophir Loop was 476 feet long and 92 feet above the canyon floor. In this scene engines Nos. 461 and 464 handle a freight from Ridgeway bound for Dolores. — ROBERT W. RICHARDSON (RIGHT) One of the famous *Galloping Goose* trains glides across a 356-foot trestle on the "high line" of the Ophir Loop region. This bridge stood 75 feet above the ground. — DONALD DUKE

California's Mount Lowe Railway was located in Southern California and was operated by the Pacific Electric Railway. It was also a bridge buff's delight. (LEFT) This photograph, taken in Millard Canyon, shows the ruggedness of this mountain electric line. The famous circular wood bridge is shown high above the track. (ABOVE) The circular bridge was visible from many points on the line. This classic view depicts the engineering marvel of the structure. — BOTH DONALD DUKE COLLECTION

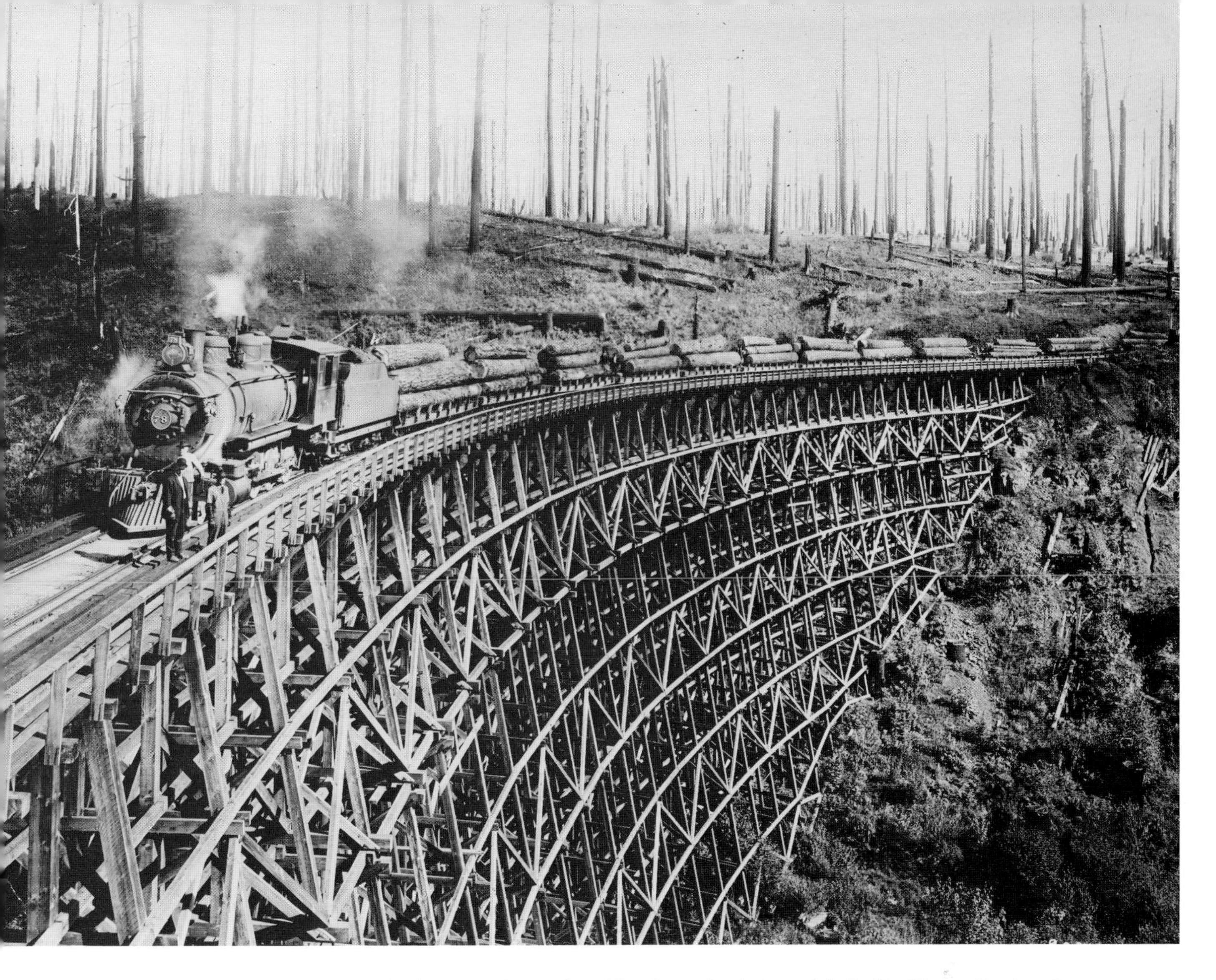

The United States Forest Service had this spectacular trestle and logging train photograph in its file. The caption is quite general and reads "Trainload of logs going to the mill in Washington State." The locomotive is No. 78 and appears to be a 2-8-0. The forest in the background has been destroyed by a major fire. — DONALD DUKE COLLECTION

California's Sugar Pine Lumber Company's line, over the years, built some 50 trestles along the 150 miles of its logging railroad. Trestle No. 14, pictured at the left, was built in 1928. The structure stood 110 feet high — the highest on the line. — AL ROSE

Northern Pacific No. 4020, a 2-8-8-2 compound, climbs the four percent grade up Lookout Pass in the Bitteroot Mountains of Idaho, near Dorsey. This line contained numerous spectacular wooden trestles. — HENRY R. GRIFFITHS

The Camas Prairie Railroad, a joint Union Pacific-Northern Pacific operation, included numerous examples of the trestle-building art on its right-of-way. In this scene, Union Pacific No. 2703 leaves Orofeno on the Headquarters line in 1954. — HENRY R. GRIFFITHS

Framed by the rugged beauty of the mountains of northern Vancouver Island, a Canadian Products Limited logging train rumbles over Rice Creek in the isolated Nimpkish Valley. More sophisticated than many wood structures, this trestle includes concrete piers and an inverted deck truss, also made of wood. — ROBERT TURNER COLLECTION

International Point Bridge
Southern Pacific

International Point — An eastbound Southern Pacific passenger train rolls out of New Mexico, and crosses the Rio Grande River into Texas. The Republic of Mexico is just outside this picture at the left. The Santa Fe rails into El Paso pass under the SP bridge. Note that the Santa Fe overcrossing and the Rio Grande River crossing have been covered. — DONALD DUKE COLLECTION (RIGHT) This same point today with a steel truss replacing the original old wood trestle. At the left, the former El Paso Southwestern line, now part of the Southern Pacific System. — RICHARD J. COOK COLLECTION

Transcontinental Railroad Bridges (Central Pacific — Union Pacific)

Few events in American history have captured the nation's imagination as did the building of the first transcontinental railroad and its completion with the driving of a golden spike into a laurel tie at Promontory, Utah, May 10, 1869. Both the Central Pacific and Union Pacific each had many trestles on their respective lines, some were large and others small — all built of wood. (ABOVE) In its haste to reach Promontory to meet the Central Pacific at the appointed time, a few hurriedly built trestles were erected four miles east of Promontory. In this scene a construction train is still at work on one of these structures. (LEFT) The Union Pacific, pushing to reach the summit of the Wasatch mountains in order to extend its rails into the Salt Lake Valley, built a zig-zag or a set of switchbacks up to the summit tunnel then under construction. In this scene the track crosses over itself. (RIGHT) All the early Central Pacific bridges were constructed of wood. The first Long Ravine Bridge, 120 feet high, was made of three Howe truss sections. The wagon road below the bridge later became the right-of-way of the Nevada County Narrow-Gauge Railroad. — ALL DONALD DUKE COLLECTION

When the Northern Pacific Railway built west into the western mountain ranges, the company did not have to go far for bridge timbers. A passenger train crosses over one of the gigantic trestles along the Coeur d'Alene line near Dorsey, Idaho, circa 1900. This photograph was taken following the rebuilding of the bridge destroyed by forest fire in 1895, evidence of which can be seen on the mountainside in the background. This structure was later replaced with a steel viaduct. — DONALD DUKE COLLECTION

The great Secret Town wood trestle, on the Central Pacific, was the longest structure of its type on the railroad. It was 1,100 feet in length and stood 90 feet in height. In order to accommodate heavier trains, the bridge was filled in shortly after it was built. — DONALD DUKE COLLECTION

Dale Creek — Union Pacific

In 1868 as the Union Pacific rails were pushing west toward Promontory, its track forces ran into a windswept battlement rising west out of Cheyenne, Wyoming. This hill, known as Sherman's Hill, became a construction and operating obstacle of unmatched proportions. Once over the top at Summit, a large gash in the earth which carried the winter runoff was experienced 14 miles east of Laramie. Known as Dale Creek, it required a major bridge in order to level the grade of the right-of-way. (LEFT) The original Dale Creek Bridge was a wood structure with 40 foot spans reaching 600 feet across, and was 120 feet above the creek. The structure, opened to traffic April 16, 1868, became very difficult to maintain, and a "Spider Web" iron bridge replaced it in 1876. It was 707 feet long and 127 feet above the creek. This bridge supported the Overland Route for nearly a decade until it too was not heavy enough for the traffic it was required to carry. (ABOVE) A passenger train crosses the "Spider Web" bridge. (LEFT) An iron girder bridge replaced the "Spider Bridge" in 1885, but was dismantled in 1901 when a line change was made, this portion of the track being abandoned, and the crossing made on a large earth and rock fill. — ALL DONALD DUKE COLLECTION

Between 1900 and 1946, the Bloedel Donovan Lumber Mills operated approximately 185 miles of logging road, in three separate operations, out of Bellingham, Washington. The line to Camp Seven had many spectacular wood trestles including this unusual Howe deck truss with trestle approach crossing over South Fork. The three-truck Climax, No. 11, represented one of the largest engines of its type built by the Climax Manufacturing Company. — DARIUS KINSEY PHOTOGRAPH FROM JESSE E. EBBERT COLLECTION

Shay geared locomotive of the Lyman Pass Railroad sses the spectacular 136-foot high wooden trestle near milton, Washington. This was the highest pile bridge in world when Darius Kinsey took this photograph in 18. The line was part of the Hamilton Logging Comny from 1907-1923. — DARIUS KINSEY PHOTOGRAPH OM JESSE E. EBERT COLLECTION

The Marent Trestle, built by the Northern Pacific on Evans Hill in 1885, 13 miles west of Missoula, Montana, was strengthened in 1945. At 226 feet, it is the highest bridge on the railroad, now a part of the Burlington Northern System. In this photograph, the westbound *North Coast Limited* crosses the 797-foot-long bridge in 1970. — RICHARD J. COOK

Boston & Maine No. 1495 handles a commute train across the high bridge at Clinton, Massachusetts, in 1955. This was near the end of operations on that section of railway line. Note the dam and powerhouse in the background. — DON ROBINSON

The Fort Dodge, Des Moines & Southern Railway crossed the Des Moines River on a high bridge just north of Boone. The structure is 800 feet long, rising 156 feet above the river and replacing a wood trestle on the same spot prior to the interurban era. The structure stands today as part of a tourist railroad operation. — ELMER R. CARR

This steel pier bridge, with wood approaches, on Pacific Electric's Glendale Line crosses over Fletcher Drive just north of downtown Los Angeles. — DONALD DUKE

Great Northern electric locomotive No. 5012 crosses Nason Creek viaduct at Old Gaynor, a few miles east of the Cascade Tunnel. — DONALD DUKE

The completion of the Virginian Railway in 1909 consummated the realization of H. H. Rogers for a super carrier from the rich bituminous mines of West Virginia to the port of Norfolk, Virginia. Like a giant conveyor belt, the Virginian moved mostly coal. Loaded trains moved generally downgrade, while lighter trains of empty coal cars tackled the heavy grades on their return. A westbound train of 12-wheel coal cars crosses the New River bridge at Glen Lyn, West Virginia, one of many viaducts on the line. The train is pulled by a 120-series electric locomotive, on 133.6 miles of electrified lines between Mullens and Roanoke, on September 4, 1953. The Virginian's entire operation was taken over by the Norfolk & Western and the old Virginian line is used primarily for eastbound trains. — RICHARD J. COOK

Typical of many plate girder bridges on the Southern Railway, this structure carries Nos. 6544 and 2166 across the recently painted bridge, that crosses over a small valley in Alabama. — JOHN KRAUSE

Akron, Canton & Youngstown's typical girder bridge takes the single-track line over Rocky River at Medina, Ohio. Lima-built No. 404, a 2-8-2, is seen here pulling a 34 car westbound mixed train, No. 95, on May 13, 1950. — RICHARD J. COOK

A view of a Chicago & North Western steel viaduct showing the construction of the supports and the piers. — DONALD DUKE COLLECTION

The second section of Great Northern Railway's prime train, the *Empire Builder,* rolls behind an oilburning S-2 class 4-8-4 engine as it takes the Gassman Coulee span into Minot, North Dakota, for St. Paul. — RICHARD J. COOK COLLECTION

Of all the scenes on the Western Pacific, perhaps the most typical and well photographed is a view of a train on the Keddie "wye" bridge. Here the last spike completing the WP was driven in 1909. In 1931 the new Bieber Line, a connection with the Great Northern, was made, thus requiring a second bridge outside a tunnel portal which formed this famous "wye" bridge. (LEFT) No. 256, a westbound freight at the "wye." (RIGHT) No. 60, a 2-8-0, heading east on the "wye" bridge. — BOTH DONALD DUKE

Old stone piers support a newer plate girder bridge on the Canadian National's line crossing the Thames River at St. Mary's, Ontario. Here, westbound train No. 111 is being pulled by 4-8-2 No. 6071 on May 17, 1958. — RICHARD J. COOK

Girder Bridges

Small metal bridges consist of a series of parallel "I" beams placed across a gap. Bridges of this type are called beam bridges, and are good for distances of about 50 feet. The number of "I" beams used depends upon the span and the load. For bridges over 50 feet, and to obtain an increase in depth, metal resembling that of "I" beams is built up from steel plates and angles. These fabricated members are called girders and if these girders are to carry track on top of them, they will more or less resemble a beam bridge. For added strength lateral and side bracing is used. The track is then placed on top of the girder flanges. (LEFT) In 1967, the Southern Pacific completed a new line over Cajon Pass. Prior to tracklaying, all the grading and bridging was completed. Here is a good view of a girder bridge under construction, showing all the framing. The scene is at Pine Lodge. — DONALD DUKE

A ponderous 2-8-8-4 Yellowstone type locomotive of the Duluth, Missabe & Iron Range, leaves Two Harbors, Minnesota, with 120 empty iron ore cars for the Range. Just out of town the rails cross over the Duluth-Two Harbors main line. The route to the Missabe Iron Range climbs 1,000 feet in less than 13 miles. Here is a good example of a plate girder bridge with steel supports. — FRANK A. KING

Providing a perfect setting to show off the grace of an eastbound Nickel Plate Berkshire No. 770, is this girder bridge over the Huron River near Avery, Ohio, photographed September 17, 1949. — RICHARD J. COOK

Chesapeake & Ohio's premier train, No. 1, the *George Washington,* is shown on the Ohio River bridge approach to the Cincinnati Union Terminal. This approach photographed, May 30, 1948, consists of a series of plate girders on supports. (LEFT) Parts of the Virginian Railway, in West Virginia, seemed to be all high bridges, many of them curved such as this one. Here electric motor No. 105 handles a westbound train at Bud, West Virginia, on June 13, 1956. — BOTH RICHARD J. COOK

Clinchfield Railroad's old, reconditioned No. 1, a handsome 4-6-0, handles a rail enthusiast train across a plate girder bridge just north of Johnson City, Tennessee. — JOHN KRAUSE

At Abbeyville, Ohio, a Baltimore & Ohio 2-8-0 No. 207 rumbles across a slender plate girder bridge supported by new concrete piers. The bridge replaced a wood trestle, typically used on the Sterling to Cleveland line. — JOHN REHOR

Having jeweled the East Coast of Florida with a string of superb hotels, Henry Flagler conceived the idea of an extension of the Florida East Coast Railway to travel over the water to Key West and a shortened steamship connection to Cuba. In this scene, the *Havana Special* travels the Long Key viaduct toward Key West. — FLAGLER MUSEUM

9

Concrete Bridges

Although the railroad engineer had not espoused so enthusiastically or as early the use of concrete as a bridge building material as did the highway engineer, still, concrete has become very acceptable for railroad use in the last half of the 20th century. Today many bridges made of reinforced concrete carry railbeds across rivers, streams, swamps or even lakes.

Concrete has taken the place of wooden piling, and concrete has replaced steel girders in many instances. Where at one time bridge piers were made of stone or steel, today they are often of concrete.

Because of the strict standards the railroads must maintain for their bridges, the possibilities of working with concrete were slow to evolve. The first reinforced concrete railroad bridges were monuments to bridge building. One of them was a memorial to one man: Henry M. Flagler, president of the Florida East Coast Railway.

Flagler's dream — which became a reality — was to build his railroad out across the Florida Everglades and across the keys to Key West. Although some said it was impossible, the huge project *was* accomplished — in concrete. The job was begun in 1903 and the railroad reached Key West, along with the first train, in 1912. This spectacular achievement saw trains operating over it for 23 years.

The Florida East Coast Railway's Overseas Extension, called the "Eighth Wonder of the World" when it was built, cost over $26 million. But its function ceased on September 2, 1935 when a 40-mile section of the line was struck by a tropical hurricane that swept some of the track and roadbed into the sea and left the rest a mass of tangled wreckage. It was then that the company decided to abandon the line completely. Eight years later, with the help of Federal funding, Flagler's "railroad that went to sea," became a highway, using many of the original railroad structures.

One of the longest bridges was Long Key viaduct, two miles long with 180 arches. The depth of the water between Long Key and

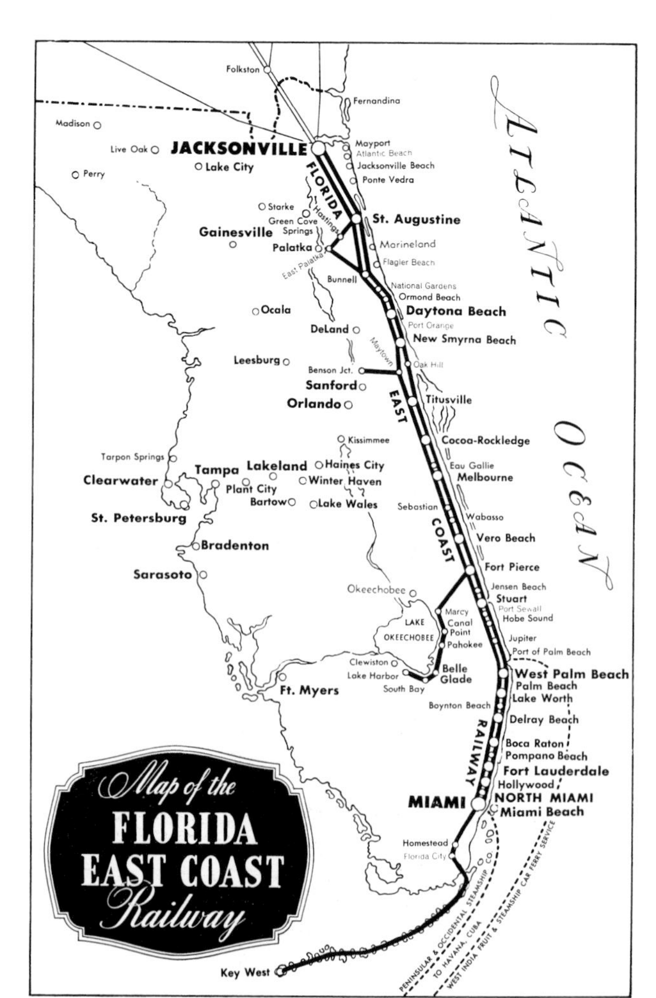

The Florida East Coast's "Overseas Extension" began with a wooden trestle at the land connection, but it soon connected with the first of many concrete bridges. The little island where the concrete spans begins is shown in the above view. (RIGHT) In the map may be seen the "Extension" of the railroad right out into the water and off to the Florida Keys. Many claimed this as "The Eighth Wonder of the World." — BOTH DONALD DUKE COLLECTION

On Long Key viaduct this American-type locomotive made easy work of its three-car train. One can only wonder how the passengers felt at viewing this seascape for such an extended period of time. The small island where the wooden trestle connected with land and the concrete spans began is shown in the background. — DONALD DUKE COLLECTION

Spanning a wide valley in graceful white arches seems to be a bit of visual poetry in this peaceful rural scene at Nicholson, Pennsylvania, where the great Tunkhannock Viaduct dominates the countryside. One can't imagine that the bridge stands 240 feet high in this view. — RICHARD J. COOK

Conch Key varied from nine to 17 feet. Heavily reinforced with steel rods, the top of the viaduct was 30 ft. above mean low water. The handling of this work, many miles from any city and far from any labor market, as well as from any supply source for materials, was, in many respects, made more difficult because of the isolation more than the physical obstacles. The work force, recruited in New York, numbered 500 to 800 men. A camp on stilts was erected on the southwest end of Long Key where the land was only slightly above water and sometimes was under water. One can only imagine what a rough and ready group it was that camped and worked together under those trying conditions. Long Key viaduct was completed in 1908. Still another bridge on the line was Knights Key bridge, 6.76 miles long, composed of both concrete arches and steel spans on concrete piers.

Another "monument" to concrete is the great and beautiful arch bridge built by the Delaware, Lackawana & Western Railroad across the half mile wide Tunkhannock Creek valley at Nicholson, Pennsylvania. This impressive concrete bridge carries two tracks of the former (now Conrail) Lackawanna main line, 2,375 feet across the valley on ten 180-foot arches rising 240 feet from the creek bed.

This awe-inspiring sight is truly one of the great bridges of America. With its symmetry, it encourages artists and photographers alike to record it. It commands the landscape and overwhelms the little town below it. In scale, it ranks with the great steel bridges of the same era, the Quebec, Sciotoville and Hell Gate.

Designed by A. Burton Cohen and built under the direction of George J. Ray, the company's chief engineer, the viaduct was the most imposing feature of a system-wide

The Tunkhannock Viaduct was the most imposing feature of a system-wide modernization program. This viaduct was the key piece in the 39.6-mile Clarks Summit-Hallstead Cut-off. The Tunkhannock Viaduct was dedicated November 6, 1915. — RICHARD J. COOK

The "No Trespassing" sign is not meant for bridge watchers who come to admire the graceful arches of the Tunkhannock Viaduct. — RICHARD J. COOK

modernization that began in the early years of the century and continued until about 1920. The program included new terminals at Hoboken and Buffalo, many new way stations, bridges and yards, double-tracking and extensive line relocations in order to reduce the grades and curves in the Pennsylvania mountains. The 39.6-mile Clarks Summit-Hallstead Cut-off and its key piece, the Tunkhannock (or Nicholson) Viaduct, were dedicated Saturday, November 6, 1915. Two special trains brought well-wishers and celebrators to the viaduct, with about 1,000 people on hand for the great day. Today, the little town of Nicholson, dominated by the great viaduct looming above, is quiet, with most of its former motor traffic bypassing it on Interstate 81. It is a rare day when Conrail runs a train across the viaduct. Another impressive concrete arch bridge on the cut-off is the 1,600-foot Martin's Creek viaduct near Kingsley, Pennsylvania.

The great Tunkhannock Viaduct overshadows the little hamlet of Nicholson, Pennsylvania, which lies at its base. — RICHARD J. COOK

Lackawanna's Martin's Creek Viaduct, also a part of the Clarks Summit-Hallstead Cut-off was eight miles east of Nicholson. — RICHARD J. COOK

Among other noteworthy concrete bridges of recent years are the Latah Creek and Indian Canyon bridges built as part of a $16.2 million Spokane line relocation on the Burlington Northern. To link the lines of the former Northern Pacific, Great Northern and the Spokane, Portland & Seattle railways, the BN engineers found that a new line was necessary. It was extended from a connection with the old main line of the NP east of Latah Creek to a connection west of the city to the old GN line to Seattle. This required construction of a long, high bridge over Latah Creek plus 5.4-miles of new trackage, including several additional bridges.

While still on the subject of great concrete bridges, it would not be right to overlook one built between 1916 and 1919. Owned jointly by the Seaboard Coast Line and the Richmond, Fredericksburg & Potomac railroads, the bridge over the James River near Richmond, Virginia, has an overall length of 2,278 feet with 15 main arch spans. It was opened June 25, 1919. Although, like the Tunkhannock Viaduct it possesses great beauty, it is only half the height of the Pennsylvania bridge and used only a quarter of the amount of building material.

The summer of 1971 saw great activity during the construction of the Spokane bridge project of the Burlington Northern. The largest of which was the Latah Creek bridge. — RICHARD J. COOK

Richmond, Fredericksburg & Potomac Railroad's bridge over the James River near Richmond, Virginia, was just as majestic as the Rappahannock River bridge at Fredericksburg, but was seldom photographed. — AUGUST A. THIEME, JR.

The Latah Creek bridge is nearly 3,700 feet long with a maximum height of about 190 feet above the stream. It includes a wye at the west end to accommodate a connection with the former Spokane, Portland & Seattle Railway line to Seattle. The bridge also carries the Union Pacific, and crosses over three important highways — Interstate 90, Inland Empire Way and Lindeke Street, as well as ramps serving I-90. The spans range in length from 70 feet to 160 feet consisting of steel-box girders with a ballasted-concrete deck. The girder design, based on continuous units of four and six spans, includes a composite deck whereby the concrete deck takes up part of the compression load through the steel studs. — UNION PACIFIC

Additional Concrete Bridges

Built by the Big Four Railroad (Cleveland, Cincinnati, Chicago & St. Louis Railroad), this concrete arch bridge over the Great Miami River at Shelby, Ohio, has seen many trains of that railroad, the Penn-Central and, now, Conrail. Heavy and fast traffic continues to use the bridge. At the time this photograph was made, a Penn-Central piggyback train crossed the bridge. — SIMON E. HERRING

These two abandoned Delaware, Lackawanna & Western Railroad bridges represent a by-gone era. Both structures are in western New Jersey. (ABOVE) This beautiful concrete bridge crosses the Delaware River at Delaware Water Gap. (LEFT) This concrete arch bridge was the Paulins Kill crossing at Hainesburg, New Jersey. — BOTH RICHARD J. COOK

The San Pedro, Los Angeles & Salt Lake Railroad (Union Pacific) crosses the Santa Ana River on a 984-foot concrete arch viaduct about five miles south of Riverside, California. The ten-arch structure is said to contain 14,000 yards of concrete and was the largest concrete bridge in the United States at the time of its completion in 1903. The viaduct consists of 10 arches flanked by two massive abutments. Eight of these arches have a clear span of 86 feet and the two end spans adjoining the abutments are about 35-foot clear spans. The piers, built of solid concrete, measure 14 feet by 28 feet. The arches are semicircular and the greatest height of the structure above the riverbed is about 70 feet. In this scene, Union Pacific's *Transcon* was photographed in 1947 crossing the viaduct en route from Los Angeles to Riverside, and points east. — DONALD DUKE

On the way from Houston to Galveston, Texas, the low marshland must be traversed by bridges like this concrete viaduct. Here Santa Fe Railway's 66-car unit train of molten sulphur, containing 858,000 gallons of the element, rolls across the viaduct near Galveston. Every three days a similar train is scheduled for the 930-mile journey from Duval Corporation's plant at Rustlers Springs, Texas, to the ocean port at Galveston. — SANTA FE RAILWAY (BELOW) This classic, solid, concrete arch bridge was completed in 1913 when the Pacific Electric built its 57-mile Los Angeles to San Bernardino line. Charles D. Savage captured car No. 1207 crossing one of the many of these concrete arch bridges on the line as it leaves San Bernardino in 1939.

The *Southern Belle* of the Kansas City Southern was photographed crossing one of the large and graceful concrete viaducts on the line near Kansas City. This concrete structure spans U.S. Highway 50 and is one of 25 overpasses within a ten-mile section of track. The streamlined *Southern Belle* provided the only direct rail passenger service between Kansas City and New Orleans. — RICHARD J. COOK COLLECTION

Since November 2, 1944, all Illinois Central trains between Paducah and Louisville have operated over the Gilbertsville Dam and bridge spanning the Tennessee River. Steam locomotive No. 2743 is a 2-10-2 with a long freight drag. — RICHARD J. COOK COLLECTION

On the maiden voyage of Western Pacific's section of the *California Zephyr,* the photographers were out in droves to capture the new train on film. One bridge buff caught the tail end of the *Zephyr* crossing the WP's concrete arch bridge in Altamont Canyon, California. — RICHARD J. COOK COLLECTION

10

Kate Shelley

Probably no story about railroad bridges would be complete without at least one anecdote or personal experience. One of the best is the story of Kate Shelley and how she saved the *Midnight Express* from crashing into the raging waters of flood-swollen Honey Creek, after the bridge had collapsed.

Kate, 15 years old, lived with her invalid mother near the main line of the Chicago & North Western at Moingona, Iowa, between Honey Creek and the Des Moines River. After the death of her father, who had been a North Western section foreman, Kate with the help of her older brother became the mainstay of the family.

On July 6, 1881, a violent rainstorm added to the swollen waters of Honey Creek, with the downpour continuing into the night. Lightning punctuated the howling wind, and thunder rattled the windows of the little Shelley house. The Shelleys were huddled inside in dry safety, reading by the light of a kerosene lamp, while outside, the entire valley was flooded. The yard of the home resembled a great oval bowl filled with water and extended to the railroad. Kate put the children to bed and then sat with her mother to wait out the storm.

Meanwhile, pusher engine No. 11, with engineer Ed Wood and crew, was out on its usual storm assignment: making a test run to check on the bridges and to scout for washouts. Midway across the Honey Creek bridge it collapsed, and the engine and structure plunged into 25 feet of rushing water.

In the Shelley home, they could hear the clanging of a bell and the sound of a "horrible crash and hissing steam" penetrating even above the noise of the storm. Kate realized instantly what had happened and, horror-stricken, snatched her father's railroad lantern from the wall and raced out into the storm. She reached the tracks and ran to the end of the wrecked bridge. In the flashes of lightning she could see two men in the middle of the swirling waters, clinging to tree limbs. One was engineer Wood, the other was

The original Des Moines River bridge was a Whipple truss on stone piers. Note that the piers appear to have been raised above the original stone foundation to allow for higher clearances over the river. — CHICAGO & NORTH WESTERN

This building is a privately operated Kate Shelley museum at Moingona, Iowa. It is not known if it was a former Chicago & North Western depot. — RICHARD J. COOK

brakeman Adam Agar. Kate knew that the *Midnight Express* was due from the west in 30 minutes and that it would crash into Honey Creek unless it was stopped at Moingona.

"Five hundred feet of bridging lay before me," Kate later explained. It was the bridge over the Des Moines River which she must cross. Doubled over by the buffeting winds and soaked to the skin, she set out to cross this 671-foot (actual length) five-truss bridge. A sudden gust of wind threw her off balance and she stumbled, causing her lantern to hit a rail and break. But nevertheless, she went on into the dark angry night without a light, making her way slowly onto the bridge, crawling on hands and knees with the rough wood and spikes tearing at her hands and clothing, using the guard rail as her guide. At every lightning flash she could see the swirling foam not far beneath her. Her heart rose in her throat when about halfway across the lightning revealed a huge uprooted tree floating in the current and bearing down on the bridge. With only this brief warning, she stopped, clung to the rail with all her strength and waited for the impact of the tree against the battered bridge. When it hit, it showered her with dirt and branches and shook the entire structure, but miraculously the bridge held. Slowly the pressure of the current forced the tree deeper into the water and finally under and free of the bridge. Kate forced herself to go on.

Ike Fansler, night operator at the Moingona station and others who had taken shelter there, were startled to see the door burst open and to hear this slim and soaking wet teenage girl cry out: "Stop the Express. Honey Creek bridge is out!" Kate — wet, disheveled, exhausted and hysterical — had made it.

To rouse a rescue crew, the whistle of the standby locomotive blasted, resulting in the waking of many of the town's citizens from their beds. Kate boarded the relief train and led the volunteers along the bluff to the washout.

Engineer Wood was thrown a rope and in a hand over hand process was able to come ashore. Agar didn't have the strength to hold onto the rope and had to remain there until daylight before he could be rescued. Foreman Patrick Donohue's body was found several days later in a cornfield, but fireman A.P.

This photograph shows Kate selling tickets as a station agent at the C&NW Moingona depot. — DONALD DUKE COLLECTION

Kate on the station platform at Moingona during the heat of summer. The station was also a water stop. — CHICAGO & NORTH WESTERN

Olmsted's body was never located.

Kate survived the storm but was ill-prepared for the torrent of fame that came afterwards. The passengers on the *Express* collected $200 for her, and her feat made headlines throughout the country. People wrote to her from all over, offering congratulations and gifts, as well as requests for souvenirs and proposals of marriage. For many years the story of her exploit was used in an Iowa school reader. Also, the Order of Railway Conductors and Brakemen presented her with an engraved pocket watch.

Memorial marker to Kate Shelley at Boone, Iowa. Placed by the Order of Railway Conductors and Brakemen on the 75th anniversary of her deed — July 6, 1956. — RICHARD J. COOK COLLECTION

After awhile Kate's life quieted down and she then tried a term of school teaching but it soon became necessary to return home to take care of her mother, brother and sisters, as hard times had fallen on the family. In the early nineties, however, a Chicago newspaper learned that Kate was going to lose her home and it raised enough money to enable her to pay off the mortgage. As a result of all of this publicity, the Kate Shelley story was revived and the State of Iowa voted her a $5,000 grant.

In 1903 Kate accepted a long-standing offer of employment from the Chicago & North Western Railway and became one of the nation's first female station agents. She worked as station agent at Moingona until her death in 1912.

Today, the line over the Des Moines River, realigned in 1901 — about four miles north of the old river crossing — uses a 2,685-foot long, 185-foot high bridge, the "Kate Shelley Bridge," to cross the river at a much higher

In 1901 the Chicago & North Western realigned its crossing of the Des Moines River, and built a new bridge about four miles north of the bridge Kate Shelley crossed. The new bridge, known to the locals as ''High Bridge,'' is 2,685 feet long and is 185 feet above the river valley. At the time this bridge was built, it was the largest double-track bridge in the world. When Kate Shelley died the structure was named the ''Kate Shelley Bridge.'' In this scene, steam locomotive No. 1024 leads a westbound passenger train across. The bridge became famous on July 30, 1986, when 16 cars of a piggyback train were blown off the bridge in a fierce rainstorm with high winds. — DONALD DUKE COLLECTION

grade level than the old one.

At the time the Boone viaduct was built, it became the largest double-track viaduct in the world. The bridge towers are 18 in number, 45 feet in width, with a maximum height of 185 feet. Stiff diagonal bracing is provided throughout and without horzontal members, except at the foot of the towers. Each of the steel verticals is made up of two 20-inch beams and one 15-inch beam. The spans are 75 feet each, with the exception of one of 300 feet, and all girders, except the larger span, have a uniform depth of seven feet with four lines of girders at this depth.

In 1956 at the 75th anniversary of Kate's feat, the C&NW named a train for her: the *Kate Shelley 400*. Although there is no longer any great drama involved with trains passing over the long bridge just west of Boone, the memory of Kate Shelley lives on in a privately operated Kate Shelley museum just outside of the small town of Moingona. In that part of the world a daring, unselfish deed is long-remembered.

Kate Shelley

The last Chicago & North Western streamliner to be inaugurated by the railroad was the *Kate Shelley 400.* The train was established during October 1955, and operated between Chicago and Boone, Iowa, a distance of 340-miles. The new train was to operate on a fast six-hour schedule, making 13 stops en route. The "400" which accompanied the name of many Chicago & North Western streamliners implied the train traveled a distance of 400 miles in 400 minutes. The consist of the *Kate Shelley 400* included coaches, diner, parlor, and baggage cars.

As part of the C&NW's belt-tightening during August 1956, train Nos. 1 and 2, the *Kate Shelley 400,* lost its ties with Miss Shelley's home territory and was cut back to Cedar Rapids, Iowa. By the fall of 1957, the train was discontinued through Iowa, and terminated at Clinton, located on the Mississippi River. As the North Western Line continued to shrink its passenger service the name *Kate Shelley 400* was discontinued in 1963. The train carrying Nos. 1 and 2 continued to operate until April 1971.

The giant steel viaduct just west of Boone, Iowa, continues to carry the name Kate Shelley Bridge.

The husky Skykomish River deck truss, on the former Great Northern Railway in Washington State, was a bridge replacement installed in 1967. The 889-foot structure is composed of a 302-foot-long deck truss, with an east approach of 163 feet, which includes two deck plate girder spans, and a west approach of 424 feet, made up of six deck plate girder spans. Maximum height of the bridge is 73 feet over the Skykomish River. — BURLINGTON NORTHERN

11

New Bridges for Old

Bridge maintenance is a continual problem, especially for railroads, where the stress on their bridges is great. When it is no longer economically feasible to make continued repairs, thought is often given toward replacing a structure. This is done either by building a completely new bridge or by making a replacement with a used bridge from somewhere else.

One railroad which made great use of replacements was the Pennsylvania Railroad. Its practice was specifically designed to save bridges for a rainy day. Whenever a change in line, track abandonment or replacement of an existing structure released a bridge from service, engineers determined if it could be reused. If reusable, the truss was carefully dismantled, all parts matched for easy reassembly, and then it was stored for future use.

Today, a span from another location can be placed on a barge and floated into place. This happened recently on the Louisville & Nashville's Chicago-Nashville-Atlanta line. An 1890 swing bridge was replaced with a used 678-foot lift bridge from Danville, Tennessee. It was floated many miles up the Tennessee river to its new location at Bridgeport, Alabama. The "new" bridge now provides a channel width of 280 feet.

In other instances, replacement bridges are installed with no interruption of vital service, as the new bridge is built alongside the old one. This was the case, for example, with the Pass Manchac bridge of the Illinois Central at Pass Manchac, Louisiana, 15 miles south of Hammond. It became necessary to replace a wooden bridge ravaged by marine worms with a concrete and steel structure. The bridge, 37 miles north of New Orleans, spans the channel that connects Lake Maurepas on the east with Lake Pontchartrain on the west. The new bridge was built just 50 feet east of the old span by means of one of the largest floating cranes in the construction industry. Mounted permanently on an 80 by 120-foot barge, the derrick had a lifting capacity of 300 tons.

This is the way that new bridges were tested: with the weight of many locomotives. Here the Central Vermont Railroad used 12 locomotives on its 1887 New Hartford bridge. — CANADIAN NATIONAL/GENE ANDERSON COLLECTION

Louisville & Nashville Railroad's relocated lift bridge at Bridgeport, Alabama, was floated up the Tennessee River from its original location at Danville, Tennessee. This bridge now provides a wide channel width of 280 feet with an overall bridge length of 678 feet. — RICHARD J. COOK

Bridge Replacement at Pass Manchac, Louisiana

On the Illinois Central's main line, at Pass Manchac, Louisiana, 15 miles south of Hammond, a new swing bridge was erected just 50 feet east of the old span in order to provide a wider channel for water traffic. While the old bridge was still in place, new concrete piers were set in place as shown in the top left. (TOP RIGHT) A huge floating crane with 300-ton capacity was brought in. The crane was used to handle all the heavy steel used in the replacement bridge. (ABOVE) The southbound *Panama Limited* was the first train over the new swing bridge. The date is January 11, 1972. The old bridge, which provided the narrower channel width, stands at the left. (LEFT) In preparation for removal of the old bridge, the rails are being removed. — ALL ILLINOIS CENTRAL GULF

Illinois Central — Cairo Bridge

The completion of the Cairo Bridge over the Ohio River on October 29, 1889, was to the history of the Illinois Central Railroad what the driving of the "Golden Spike" was to the completion of the transcontinental railroad. The bridge signaled the completion of the last link in a great chain of rail communication forming a direct route between the Great Lakes and the Gulf of Mexico. The old 1889 nine Whipple truss span bridge was replaced with this six Warren truss bridge in 1951. — RICHARD J. COOK

Illinois Central's venerable crossing of the Ohio River, at Cairo, Illinois, was a nine span Whipple through truss bridge which had put in many years of faithful service. The 1889 vintage bridge became a hazard to river navigation as towboats were unable to pass adequately under the bridge and make a turn in the river. At the request of the United States Coast Guard and other river navigation authorities, the bridge was replaced with a six span structure built on the existing piers. — ILLINOIS CENTRAL GULF

This sequence picture story shows the old bridge being replaced by the new spans. The old bridge span was shifted to one side and sent plunging into the river below. The new span was then pushed into place. It took a day per span to replace. The new bridge, consisting of six Warren through spans and six deck truss spans, is a vital link in the north-south traffic of Illinois Central Gulf and Amtrak. — ILLINOIS CENTRAL GULF

The bridge rests on 100-foot piles driven into the bottom of the channel through which nearly 4,000 boats pass annually. The old bridge was a swing span. The new one is a single-leaf bascule which gives an 85-foot clear channel. The movable span is a 121-foot through truss; the entire length of the bridge is 2,484 feet.

One of the most dramatic bridge replacements — and this on the same railroad — was the removal of the old Illinois Central main line nine-truss bridge over the Ohio River at Cairo, Illinois. After the new bridge was completed alongside the old one, the now "redundant" bridge was dynamited into the murky Ohio River waters, falling span-by-span onto the banks or into the water below.

Structural fatigue, an ailment that afflicts all bridges, implies the failure of a member or connection due to the repeated application of a stress less than that which could cause failure if applied only once. In a railroad bridge, repeated stress applications result from locomotives and cars crossing the bridge. Over the years, a bridge may have carried a large number of trains and, of course, the older the bridge the more stress applications it has had and the more likely it may become a fatigue factor. Structural fatigue is an old-age disease, just as in humans.

Fatigue cracks will most likely be found where there is a high stress concentration such as at an abrupt change in geometry. These changes will be found at reentrant cuts, holes and notches or at abrupt changes in stiffness.

Even though the pounding inflicted by steam locomotives is now gone, over a long period of time, the rhythmic, moving stress of passing long, heavy freights pulled by several heavy diesel locomotives still results in considerable damage to bridges.

Railroad bridges have been in the general news during the past few years because of accidents or failures. These conditions have been rectified and, it is hoped, that better maintenance will result in fewer accidents in the future.

In the Boston area, Boston & Maine and the Massachusetts Bay Transportation Authority suffered a great loss of passenger revenues when wood pilings of the twin bascule bridges

Two additional replacement bridges for the Illinois Central Gulf were built over the Little Calumet River at Riverdale, Illinois, a Chicago suburb. The elimination of a center pier, which proved a hazard to navigation, was the reason for the two adjacent electric spans and a single track non-electrified bridge begun May 1, 1969, and completed in August 1972. During most of the construction work on the $9.5 million project, three of the five tracks were kept open at all times for passenger, freight, and suburban electric traffic. The three spans now provide a 300-foot clearance for barge traffic on the river. (ABOVE) A temporary bridge was built alongside the new non-electrified span until it was connected in place. (BELOW) One of the new electrified bridges and the non-electric bypass bridge. — BOTH RICHARD J. COOK

A suburban electric "Highliner" whisks across the completed project. Note the non-electrified bridge in the background. — RICHARD J. COOK

The first and third bridge across Mountain Creek on the Canadian Pacific in the Selkirk Mountains are shown here. (ABOVE) The first bridge, built in 1885, was a sturdy wood trestle. In time it became hard to maintain and could not support heavier trains. (BELOW) A girder is put into place on a new $3 million bridge completed in 1978. At 3,000 feet above sea level, more than 600 tons of steel and 160,000 tons of fill were employed to complete the project. The new structure has a concrete ballasted deck for more efficient maintenance and the elimination of fire hazard. The second bridge, a steel viaduct which was in use while the new bridge was under construction, may be seen in the background. —
BOTH CANADIAN PACIFIC

over the Charles River caught fire and were destroyed, resulting in a blocked entrance and exit to North Station. A temporary station on the west side of the river was set up and the commuter operation dead-ended there until the pilings (this time made of concrete) could be restored.

Another Boston & Maine bridge fire, in the electrical box of the Danvers drawbridge kept the 98-year old bridge and trestle out of service (commuter service) for many months.

In the midwest, Illinois Central Gulf's Mississippi River bridge at Dubuque, Iowa, became locked in the raised position on October 6, 1984. It stranded the 13-car Amtrak Galena-Dubuque Limited excursion train for six hours. The International lift bridge linking the Soo Line and CP Rail at Sault Ste. Marie, Michigan-Ontario became tilted at a 15-degree angle, in a half-raised position. For two days it stopped ship traffic and involved 18 boats before the problem was corrected.

In Philadelphia, the SEPTA/Reading bridge at Temple University station was condemned on very short notice and ordered dismantled, forcing painful rerouting of trains. The bridge has since been rebuilt.

Brand new bridges are also being erected today, most of them in the West. On CP Rail, the New Rogers Pass line in the Canadian Rockies has required a replacement of the Mountain Creek bridge at 3,000 feet above sea level. The two-year project resulted in a 600-foot girder bridge 136 feet above Mountain Creek, resting on two steel towers and two concrete piers. It is the third bridge over Mountain Creek since the opening of the Canadian Pacific line in 1885. The new bridge replaces a 1902 585-foot girder span, strengthened in 1928-29. The first bridge was a higher (164 feet) and longer (1,086 feet) timber trestle containing more than two million board feet of lumber. In its day, it was a magnificent bridge.

At Index, Washington, the Great Northern Railway (now Burlington Northern) found the need for a heavier bridge over the Skykomish River and, in 1967, completed a 889-foot crossing with a 302-foot deck truss span. The bridge is supported by concrete piers and abutments. The substructure units are founded either directly on bedrock or rest on steel piles driven into bedrock.

Structural fatigue, an ailment that affects all bridges, affected the old Nimrod bridge of the Great Northern near Essex, Montana. The new structure was open to traffic February 28, 1966, and was photographed during a winter storm. — BURLINGTON NORTHERN

New concrete piers replaced steel piers on the Canadian National bridge over Twelve Mile River at St. Catherines, Ontario. A train of RDC (Rail Diesel Cars) heads west to Hamilton from Niagara Falls and crosses over the 100-foot deep valley. — RICHARD J. COOK

Still the work goes forward, resulting in heavier and more useful bridges which help to shorten schedule times, to reduce grade factors, to sustain longer and heavier freights and, in the long run, to help the railroad and their customers save money. It will be interesting to see, in the years to come, what the next great bridges on our railroads will be.

Perhaps they will be typified by the new line that Burlington Northern built at Spokane, Washington, extending from a connection with the old main line of the Northern Pacific east of Latah Creek to a connection west of the city to the old Great Northern line to Seattle. The $16.2 million project required construction of a long high bridge over Latah Creek plus 5.4 miles of new trackage, including several additional bridges.

The Latah Creek bridge is nearly 3,700 feet long with a maximum height of about 190 feet above the stream. At the west end, it includes a wye to accommodate a connection with the

The summer of 1971 saw great activity in the construction of the Burlington Northern's Spokane bridge project, the largest structure of which was the Latah Creek bridge. — RICHARD J. COOK

A bridge worker walks on one of the huge girders on a section of the new Latah Creek bridge just west of Spokane. Today, the 16.2 million Burlington Northern project sees heavy freight traffic. — RICHARD J. COOK

former SP&S line to Seattle. The bridge carries the track over the Union Pacific and three important highways. The superstructure consists of steel-box girders with a ballasted concrete deck. The girder design, based on units of four and six spans, includes a composite deck whereby the concrete deck takes part of the compression load through steel studs. The girders are supported on rectangular reinforced-concrete piers. The project was begun early in 1971 and was completed in December 1972. The concrete piers lend an airy, graceful look to the new bridges.

Tomorrow's bridge engineers — men of vision — will, like their predecessors, be pioneers, conquering the many obstacles confronting them and leaving their mark for future generations to appreciate.

A Burlington Northern freight train crosses the Latah Creek bridge at Spokane, Washington. At one point on the concrete viaduct, the tracks rise to a height of 205 feet above the ground. In order to check the bridge, inspectors need only open a trap door, enter the steel "box girder" which supports the tracks, then walk through the lighted interior. — BURLINGTON NORTHERN

WROUGHT-IRON BRIDGE CO.

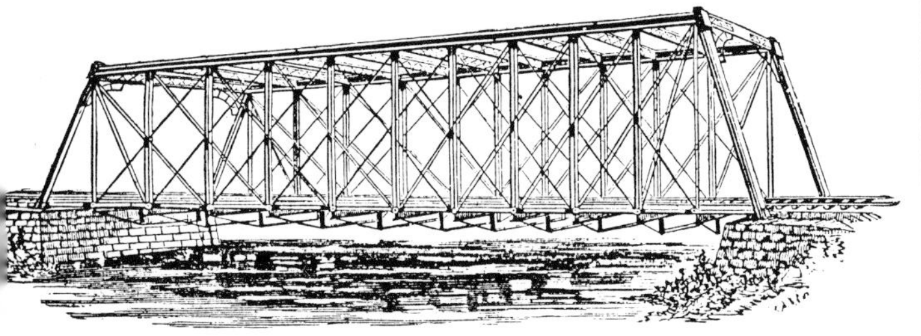

AMMOND, President. | Office and Works, CANTON, O. | JOB ABBOTT, H. G. MORSE, } Engineers.
BRITTON, Secretary.

Manufacturers and Builders of all Wrought-Iron Railway and Highway Truss, Arch and Swing ges, Plate and Lattice Girders, Iron Roofs, Turn-Tables, Iron Piers and Trestles. Have over NTY-FOUR MILES of their IRON BRIDGES now in use in twenty-four different States and da. ILLUSTRATED ALBUM and estimates sent on application.

OUISVILLE BRIDGE & IRON CO.,

Office and Works: Corner Oldham and 11th Streets, Louisville.

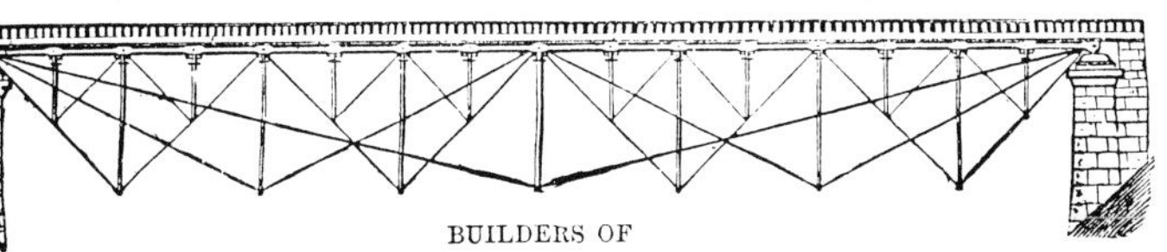

BUILDERS OF

FINK'S SUSPENSION AND TRIANGULAR TRUSSES,

other forms of Iron and Combination Bridges. Also Manufacturers of Iron Roofs, Turn-tables, s, Switches, etc.

ALBERT FINK, President.
A. P. CHROCAN, Secretary.
F. W. VAUGHAN, GILMAN TRAFTON, } Engineers
E. BENJAMIN, Superintendent of Works.

ING IRON BRIDGE & MANUFACTURING CO.,

CLEVELAND, OHIO.

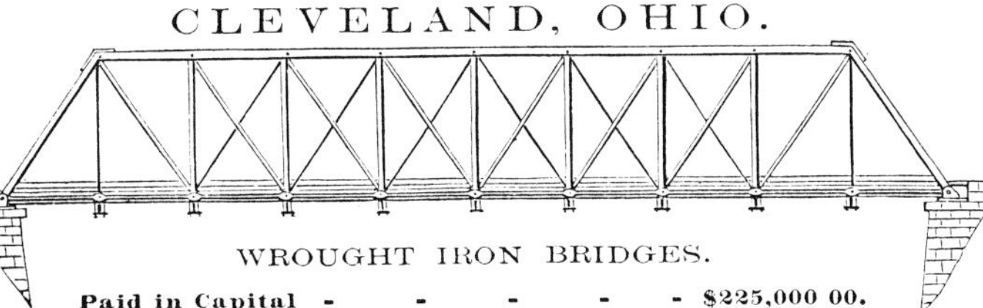

WROUGHT IRON BRIDGES.

Paid in Capital - - - - - - $225,000 00.

THE AMERICAN BRIDGE CO.

A. B. STONE, 20 Nassau st., N. Y., Pres. | EDWARD HEMBERLE, } Engineers.
H. A. RUST, Vice-Pres't & Gen'l Manager. | W. G. COOLIDGE, Sec'y.

MANUFACTURERS AND BUILDERS OF

BRIDGES,

oofs, Turning-Tables, Pivot Bridges, Iron Trestles Wrought Iron Columns Heavy Castings,

GENERAL IRON AND FOUNDRY WORK.

BUILDERS OF

PNEUMATIC, MASONRY, AND SCREW-PILE SUBSTRUCTURES.

Iron Bridges and Roofs upon the principal Railroads in the United States illustrate designs and ent the character and extent of products of Works.

☞ Proposals accompanied by Plans, Specifications and Lithographs promptly submitted upon plication.

ORKS: Cor. Egan and Stewart Aves. | OFFICE: No. 210 LaSalle st., Cor. Adams.

Address THE AMERICAN BRIDGE CO., Chicago.

BALTIMORE BRIDGE CO.,

54 Lexington Street, Baltimore.

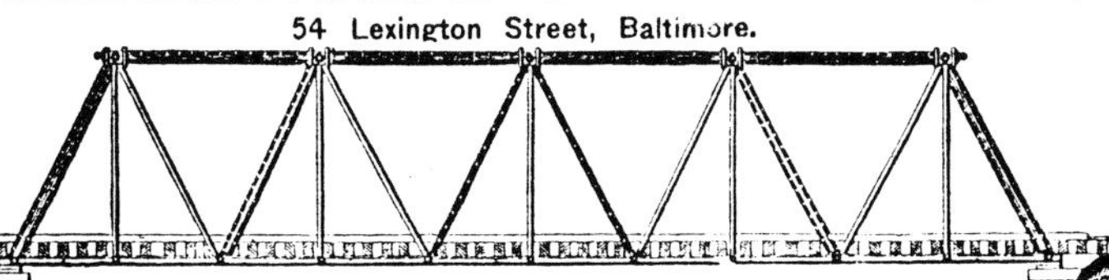

C. SHALER SMITH, Prest. & Ch. Eng.,
C. C. WRENSHALL, Supt. of Erection,
C. H. LATROBE, A. Eng., Sec. & Treas.,
C. W. BAYLEY, Superintendent,

WESTERN OFFICE, 417 Pine Street, St. Louis.
EASTERN OFFICE, 54 Lexington Street, Baltimore.

F. H. SMITH, A. Eng. & Gen. Supt., Construction Office, Wilmington, Del.

esign and Construct Iron, Steel and Composite Bridges and Roofs of any form or span desired. pecial attention is called to our Wrought-Iron Trestles and Viaducts, patented in United States, gland, France and Belgium, and built of any height or length. ithographs and information promptly furnished.

NIAGARA BRIDGE WORKS,

BUFFALO, N. Y.,

BUILDERS OF IRON LATTICE BRIDGES

FOR RAILROADS AND HIGHWAYS.

PLATE GIRDERS, COMPOSITE BEAMS, ROOFS, PLATE TURNTABLES, &c.

S. J. FIELDS, ENGINEER. GEO. C. BELL, MANAGER.

Office and Works: Forest avenue, cor. Niagara street.

Appendix

THE MILWAUKEE BRIDGE AND IRON WORKS,

G. W. JAMES & CO., Milwaukee, Wis.

BALTIMORE & OHIO RAILROAD BRIDGE, SOUTH CHICAGO, ILL.

THE KEYSTONE BRIDGE COMP'Y,

PITTSBURGH, PA.,

Engineers and Builders of WROUGHT-IRON and WOODEN BRIDGES, IRON ROOFS and BUILDINGS
Manufacturers of Wrought-Iron Turntables, Wrought-Iron Columns, Weldless Chords, Castings, Rolls, General Machine and Mill Work. Illustrated Album sent on application to undersigned.

J. H. LINVILLE, President,
218 SOUTH-FOURTH STREET, Philadelphia.

The Watson Manufacturing Company.

PATERSON, NEW JERSEY,

BUILDERS of

Post's Patent Diagonal Truss Bridges,

OF IRON, WOOD, OR WOOD AND IRON COMBINED, FOR

Railroads and Highways, Roofs, Turn-Tables and General Machine Work.

NEW YORK OFFICE, 303 BROADWAY.

W. G. WATSON, Pres. J. WATSON, Vice-Pres. A. J. POST, Engineer.

Bibliography

BOOKS

Alexander, Edwin P. *Model Railroads: Planning-Construction-Operation.* New York: W.W. Norton & Co. Inc., 1940.

Allen, Richard Sanders *Early American Bridges.* Brattleboro: Stephen Greene Press, 1960.

Ames, Charles Edgar *Pioneering the Union Pacific.* New York: Appleton-Century-Crofts, 1969.

Andrews, Ralph W. *Timber.* Seattle: Superior Publishing Co., 1968.

Arnold, Edward *Design of Steel Bridges.* New York: McGraw-Hill Book Co., 1915.

Beaver Roy C. *The Bessemer & Lake Erie Railroad.* San Marino: Golden West Books, 1969.

Berg, Walter C. *Buildings & Structures of American Railroads.* New York: John Wiley & Sons, 1893.

Best, Gerald M. *Iron Horses to Promontory.* San Marino: Golden West Books, 1969.

Bryant, Keith L., Jr. *History of the Atchison, Topeka & Santa Fe Railway.* New York: Macmillan Publishing Co. Inc., 1974.

Bryant, Ralph Clement *Logging Railroad Construction.* New York: John Wiley & Sons, Inc., 1914.

Condit, Carl W. *American Building Art.* New York: Oxford University Press, Inc., 1961.

Dredge, James *The Pennsylvania Railroad: It's Organization, Construction, and Management.* New York: John Wiley & Sons, 1879.

Due, John F. and Juris, Francis *Rails to the Ochoco Country.* San Marino: Golden West Books, 1968.

Hofsommer, Donald L. *The Southern Pacific 1901-1985.* College Station: Texas A&M University Press, 1986.

Hool, George Albert *Movable and Long Span Steel Bridges.* New York: McGraw-Hill, 1923.

Holland, Rupert Sargent *Big Bridge.* Philadelphia: Macrae-Smith Co., 1938.

Hopkins, H.J. *A Span of Bridges.* Newton Abbott: David & Charles, 1970.

Jacobs and Neville *Bridges, Canals and Tunnels.* New York: American Heritage Publishing Co., 1968.

Jameson, Charles Davis *The Evolution of the Modern Railway Bridge.* Iowa City: University of Iowa Press, 1916.

Kratville, William W. *Golden Rails.* Omaha: Kratville Publications, 1965.

Krause, John *Rails Through Dixie.* San Marino: Golden West Books, 1965.

McCullough, Conde B. *Elastic Arch Bridge.* New York: John Wiley & Sons, Inc., 1931.

McCullough, David *The Great Bridge.* New York: Simon & Schuster, 1972.

McLean, Harold H. *Pittsburgh & Lake Erie Railroad.* San Marino: Golden West Books, 1969.

Mallery, Paul *Bridge and Trestle Handbook.* New York: Simmons-Boardman Publishing Co., 1958.

Modjeski & Masters *Final Report on Reconstruction - Illinois Central Cairo Bridge Over the Ohio.* Mechanicsville: Modjeski & Masters, 1947.

Plowden, David *Bridges, The Spans of North America.* New York: The Viking Press, 1974.

Reid, H. *The Virginian Railway.* Milwaukee: Kalmbach Books, 1961.

Scribbins, Jim *The 400 Story.* Park Forest: PTJ Publishing, 1982.

Steinman, David B. *Builders of the Bridge.* New York: Harcourt, Brace & Co., 1945.

Steinman, David B. *Famous Bridges of the World.* New York: Random House, 1953.

Steinman, David B. and Watson, Sara Ruth *Bridges and Their Builders.* New York: G.P. Putnam's Sons, 1941.

Tratman, E.E. Russell *Railway Track & Structures.* New York: McGraw-Hill Book Co., 1909.

Waddell, J.A.L. *Bridge Engineering.* New York: John Wiley & Sons, Inc. 1916.

Watson, Wilbur Jay *A Decade of Bridges.* Cleveland: H.J. Janson, 1937.

Whitney, Charles S. *Bridges: Their Art, Science & Evolution.* New York: Greenwich House, 1983.

Abbey, Wallace W. "Hardest Thing I Had to Do." *Trains,* Vol. 14, December 1953: pp. 48-49.

Abbey, Wallace W. "I Flew Across the Hudson on a Freight Train." *Trains,* Vol. 13, May 1953: pp. 26-27.

Abbey, Wallace W. "Keddie Wye - Almost!" *Trains,* Vol. 13, March 1955: pp. 48-49.

Abbey, Wallace W. "Samson of the Cimarron." *Trains,* Vol. 13, February 1953: pp. 48-49.

Abbey, Wallace W. "They Called It Latrobe's Folly." *Trains,* Vol. 13, December 1952: pp. 20-21.

Ackerman, John H. "Across Buzzard's Bay." *Trains,* Vol. 16, October 1955: pp. 56-58.

Allen, Richard Sanders "Crossing Under Cover." *Trains,* Vol. 15, June 1955: pp. 44-50.

Allen, Richard Sanders "Over Niagara Falls on a Wire." *Trains,* Vol. 18, December 1957: pp. 44-47.

Anonymous "America's Debt to the Howe Truss Timber Railway Bridge." *Scientific American,* May 1922: p. 90.

Anonymous "Concrete Railroad Bridge at Danville, Illinois." *Scientific American Supplement,* August 11, 1906: p. 520.

Anonymous "Continuous Trusses of Silicon Steel Feature of New Allegheny Bridge." *Engineering News-Record,* Vol. 80 No. 18: pp. 848-856.

Anonymous "Double-Track Railway Viaduct Over the Des Moines River." *Scientific American,* June 1, 1901: p. 340.

Anonymous "James River Bridge." *Scientific American Supplement,* October 13, 1917: p. 840.

Anonymous "The Cantilever Railroad Bridge Across Niagara Gorge." *Scientific American Supplement,* March 3, 1908: p. 202.

Anonymous "The Fourth Highest Bridge in the World." *Southern Pacific Bulletin* (Texas and Louisiana Lines), Vol. 8 No. 11, November 1922: pp. 1-2.

Banks, Jimmy "Paragon of the Pecos." *Railway Progress,* Vol. 6 No. 9, November 1952: pp. 40-53.

Boomer, Pete "Bridge Engineering." *Model Railroader,* March 1944: pp. 120-122.

Boomer, Pete "Bridges." *Model Railroader,* December 1946: pp. 828-831.

Bowman, Fielding L. "Hell Gate." *Trains,* Vol. 14, October 1954: pp. 27-30.

Chase, C.E. "The Crooked River Arch." *Scientific American Supplement,* February 7, 1914, pp. 92-93.

Ellison, Frank C. "Deck Truss Bridge." *Model Railroader,* July 1940: pp. 369-371.

Fisher, Charles E. "Niagara Falls Suspension Bridge." *Railway & Locomotive Historical Society Bulletin No. 23,* November 1930: pp. 56-57.

Fowle, F.P. "The Original Rock Island Bridge Across the Mississippi River." *Railway & Locomotive historical Society Bulletin No. 56,* October 1941: pp. 55-63.

Galvin, E.D. "The Canton Viaduct." *Railway & Locomotive Historical Society Bulletin No. 129,* Autumn 1973: pp. 71-85.

Gross, H.H. "Steel Across the River - Part One." *Railroad Magazine,* Vol. 44 No. 4, January 1948: pp. 8-38.

Gross, H.H. "Steel Across the River - Part Two." *Railroad Magazine,* Vol. 45 No. 1, February 1948: pp. 10-27.

Gross, H.H. "The Pecos Legend." *Railroad Magazine,* Vol. 49, No. 1, July 1949: pp. 42-57.

Houghton, Richard "Masonry Bridges." *Model Railroader,* April 1948: pp. 253-254.

Houghton, Richard "Trestle Construction." *Model Railroader,* February 1948: pp. 96-97.

Houghton, Richard "Timber Trestles." *Model Railroader,* February 1946: pp. 82-85.

Hubbard, Freeman "The Great Starucca Viaduct." *Railroad Magazine,* Vol. 39 No. 3, February 1946: pp. 106-107.

Kalmbach, A.C. "Bridge Abutments." *Model Railroader,* December 1937: pp. 460-461.

Klug, Harry "Canadian National Bridges." *Trains,* Vol. 1, Janaury 1941: pp. 14-18.

McLaughlin, D.W. "Poughkeepsie Gateway." *Railway & Locomotive Historical Society Bulletin No. 119,* October 1968: pp. 6-33.

Madle, Alain "Double-Span Girder Bridges." *Model Railroader,* September 1942: pp. 406-410.

Mallery, Paul "Skewed Deck Truss Bridges." *Model Railroader,* February 1958: pp. 54-57.

Parker, George A. "The Susquehanna Bridge on the Philadelphia, Wilmington and Baltimore Railroad." *Railroad History,* Spring 1976: pp. 39-55.

Schmeizer, Ken "Wooden Bridges and Railroad Loggers." *Bulletin* - National Model Railroad Association, Vol. 39 No. 3 Issue 390, November 1973: pp. 50-53, 59.

Ramsdell, Roger C. "The Covered Bridge." *Model Railroader,* Janaury 1947: pp. 8-9.

Smith, Ken "Modern Steel Truss Bridge." *Model Railroader,* March 1961: pp. 40-45.

Stevens, Erie "Bridges." *Model Railroader,* June 1954: pp. 58-74.

Taylor, Frank "A Timber Trestle." *model Railroader,* May 1940: pp. 252-255.

Taylor, Frank "Plate Girder Bridges." *Model Railroader,* March 1943: pp. 108-111.

Taylor, Frank "Truss Bridges." *Model Railroader,* February 1957: pp. 18-19.

Taylor, Frank "Steel Truss Bridge." *Model Railroader,* December 1942: pp. 538-543.

Vogal, Robert M. "Eads Bridge." *Railway & Locomotive Historical Society Bulletin No. 124,* April 1971: p. 34.

Vogal, Robert M. "Thomas Viaduct." *Railway & Locomotive Historical Society Bulletin No. 124,* April 1971: p. 35.

Wilson, Oliver Whitwell "A Modern Arch Bridge." *Model Railroader,* March 1936: pp. 61-63.

Wilson, Oliver Whitwell "Plate Girder Bridges." *Model Railroader,* November 1938: pp. 485-486.

Work, Jack "Wood Truss Bridge." *Model Railroader,* April 1960: pp. 27-31.

Wright, Malcom K. "The Southern Pacific Co's. New Martinez-Benicia Bridge." *Baldwin Locomotives,* Vol. 9 No. 4, April 1931: pp. 31-44.

DOCUMENTS

American Soceity of Civil Engineers, Proceedings, Volume 94.

Congressional Record

Second Annual Report, San Francisco-Oakland Bay Bridge, 1936.

TECHNICAL AND TRADE JOURNALS

Civil Engineering, New York, 1930-1980

Engineering News, London, 1961-1980

Engineering News-Record, New York, 1874-1980

Mechanical Engineering (American Society of Mechanical Engineers), New York, 1906-1980

Railway Age, New York, 1972-1980

Railway Age Gazette, New York, 1892-1915

Railway Track and Structures, New York, 1885-1910

Scientific American, New York, 1880-1952

Index

* - Indicates Photograph or Illustration

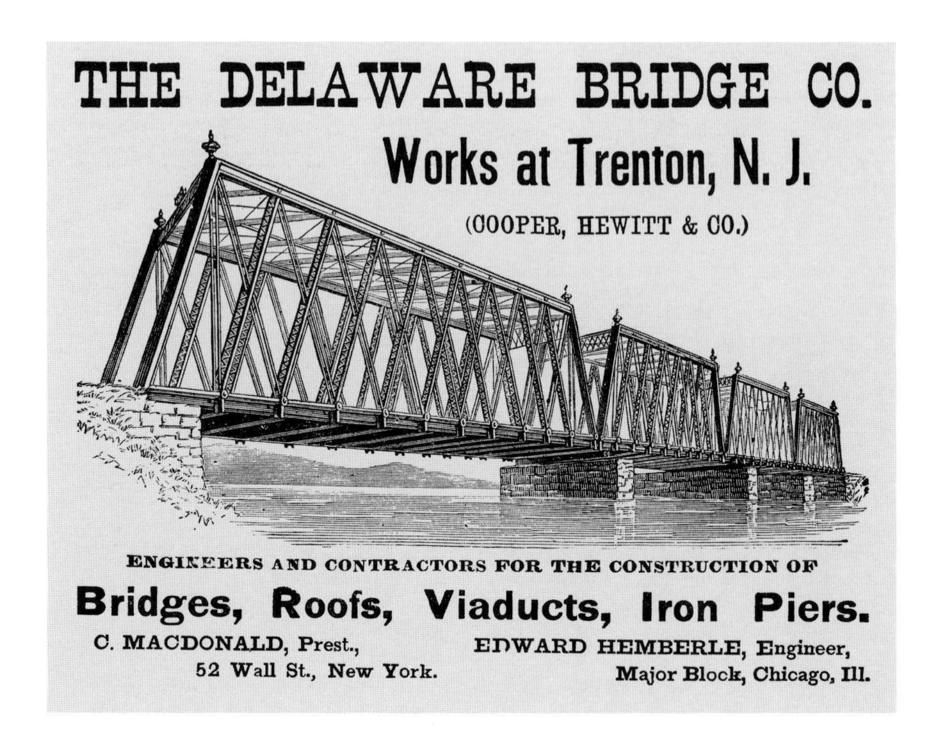
THE DELAWARE BRIDGE CO.
Works at Trenton, N. J.
(COOPER, HEWITT & CO.)
ENGINEERS AND CONTRACTORS FOR THE CONSTRUCTION OF
Bridges, Roofs, Viaducts, Iron Piers.
C. MACDONALD, Prest.,
52 Wall St., New York.
EDWARD HEMBERLE, Engineer,
Major Block, Chicago, Ill.

TIMBER TRESTLE BRIDGE

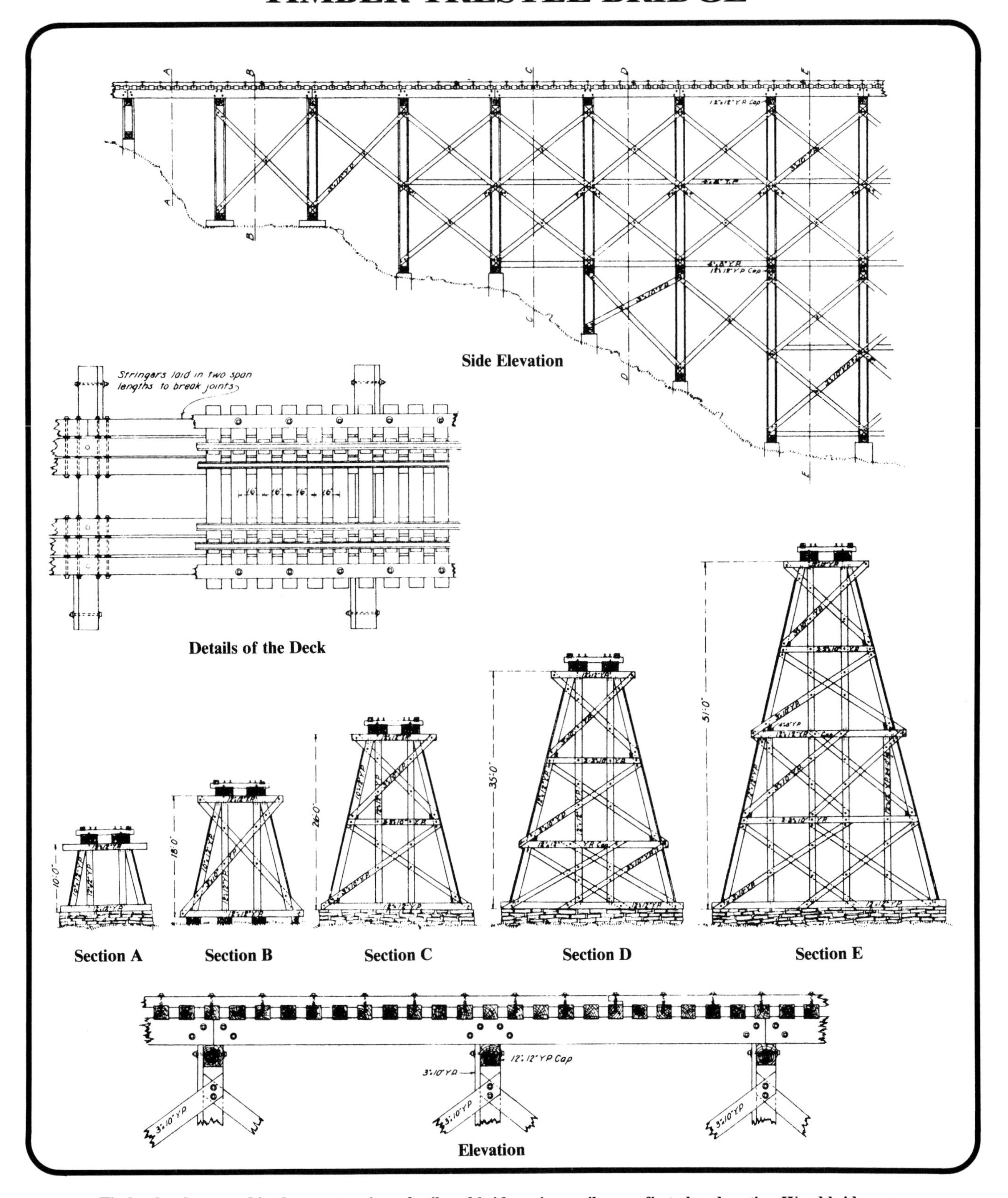

Side Elevation

Details of the Deck

Section A

Section B

Section C

Section D

Section E

Elevation

Timber has been used in the construction of railroad bridges since rails were first placed on ties. Wood bridges are built of two types, pile and cut framed timber, and then erected in sections, called bents that are spaced 12 to 14 feet apart. The length of the bents depend on the contour of the ground. While steel and reinforced concrete have replaced many of the large timber trestles, timber is still used for small creeks and river crossings. This Pennsylvania Railroad standard timber trestle plan shows the side elevation and how the bent sections are placed according to the height of the structure.

HOWE TRUSS BRIDGE

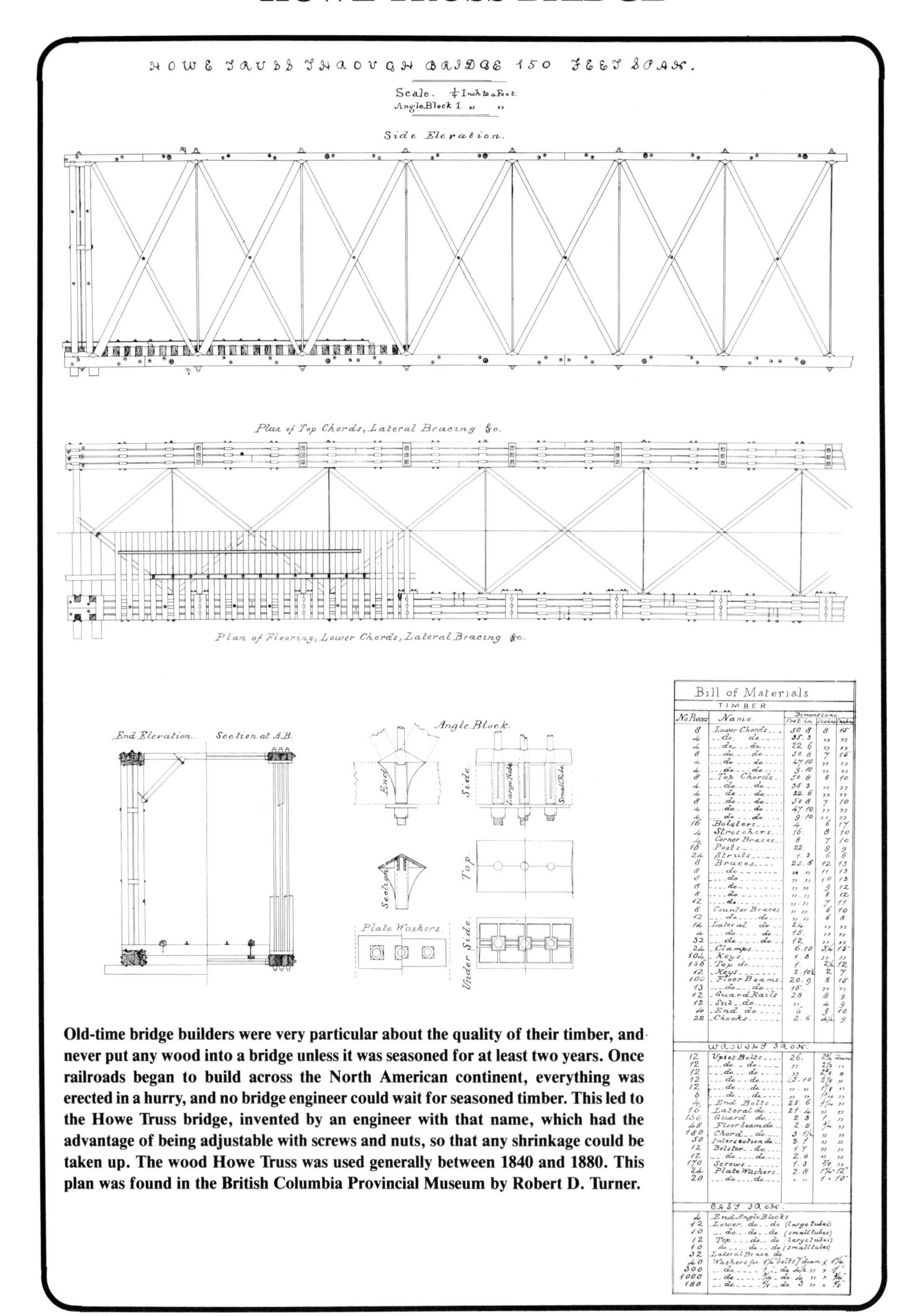

Bill of Materials

No. Pieces	Name.	Dimensions. Feet in	Inches	Inches
	TIMBER.			
8	Lower Chords	50. 8	8	15
4	do. do.	35. 3	"	"
4	do. do.	22. 6	"	"
8	do. do.	50. 8	7	15
4	do. do.	47. 10	"	"
4	do. do.	9. 10	"	"
8	Top Chords	50. 8	8	10
4	do. do.	35. 3	"	"
4	do. do.	22. 6	"	"
8	do. do.	50. 8	7	10
4	do. do.	47. 10	"	"
4	do. do.	9. 10	"	"
16	Bolsters	4.	6	17
4	Stretchers	16.	8	10
4	Corner Braces	8	7	10
16	Posts	22.	9	9
24	Struts	1. 3	6	6
8	Braces	25. 6	12	13
8	do	" "	11	13
8	do	" "	10	13
8	do	" "	9	12
8	do	" "	8	12
12	do	" "	7	11
8	Counter Braces	" "	6	10
12	do. do.	" "	6	8
14	Lateral do.	24.	"	"
4	do. do.	15.	"	"
32	do. do.	12.	"	"
24	Clamps	6. 10	3¼	15
104	Keys	1. 3	"	"
156	Top do.	1.	2½	12
12	Keys	2. 10½	2	7
100	Floor Beams	20. 9	8	15
13	do. do.	15.	"	"
12	Guard Rails	28	8	9
12	Sub do.	"	4	9
4	End do.	6	9	10
22	Chocks	2. 6	4½	9
	WROUGHT IRON.			
12	Upset Bolts	26.	2¾	diam
12	do. do.	"	2½	"
12	do. do.	"	2⅜	"
12	do. do.	25. 10	2⅛	"
12	do. do.	" "	1⅞	"
6	do. do.	" "	1¾	"
4	End Bolts	25. 6	1¼	"
16	Lateral do.	21. 4	"	"
150	Guard do.	2. 3	1	"
48	Floor beam do.	2. 8	¾	"
180	Chord do.	3. 1½	"	"
50	Intersection do.	3. 1	"	"
12	Bolster do.	1. 7	"	"
12	do. do.	2. 0	"	"
170	Screws	1. 3	⅝	"
24	Plate Washers	2. 8	1¼ × 12"	
20	do. do.	" "	1 × 10"	
	CAST IRON.			
4	End Angle Blocks			
12	Lower do. do (large tubes)			
10	do. do. do (small tubes)			
12	Top do. do (large tubes)			
10	do. do. do (small tubes)			
32	Lateral Brace do			
40	Washers for 1¼" bolts 7" diam × 1¼"			
300	do. 1" do 4½ " × 1"			
1000	do. ¾ do. 4 " × ¾"			
180	do. ⅝ do. 3 " × ⅝"			

Old-time bridge builders were very particular about the quality of their timber, and never put any wood into a bridge unless it was seasoned for at least two years. Once railroads began to build across the North American continent, everything was erected in a hurry, and no bridge engineer could wait for seasoned timber. This led to the Howe Truss bridge, invented by an engineer with that name, which had the advantage of being adjustable with screws and nuts, so that any shrinkage could be taken up. The wood Howe Truss was used generally between 1840 and 1880. This plan was found in the British Columbia Provincial Museum by Robert D. Turner.

SWING BRIDGE

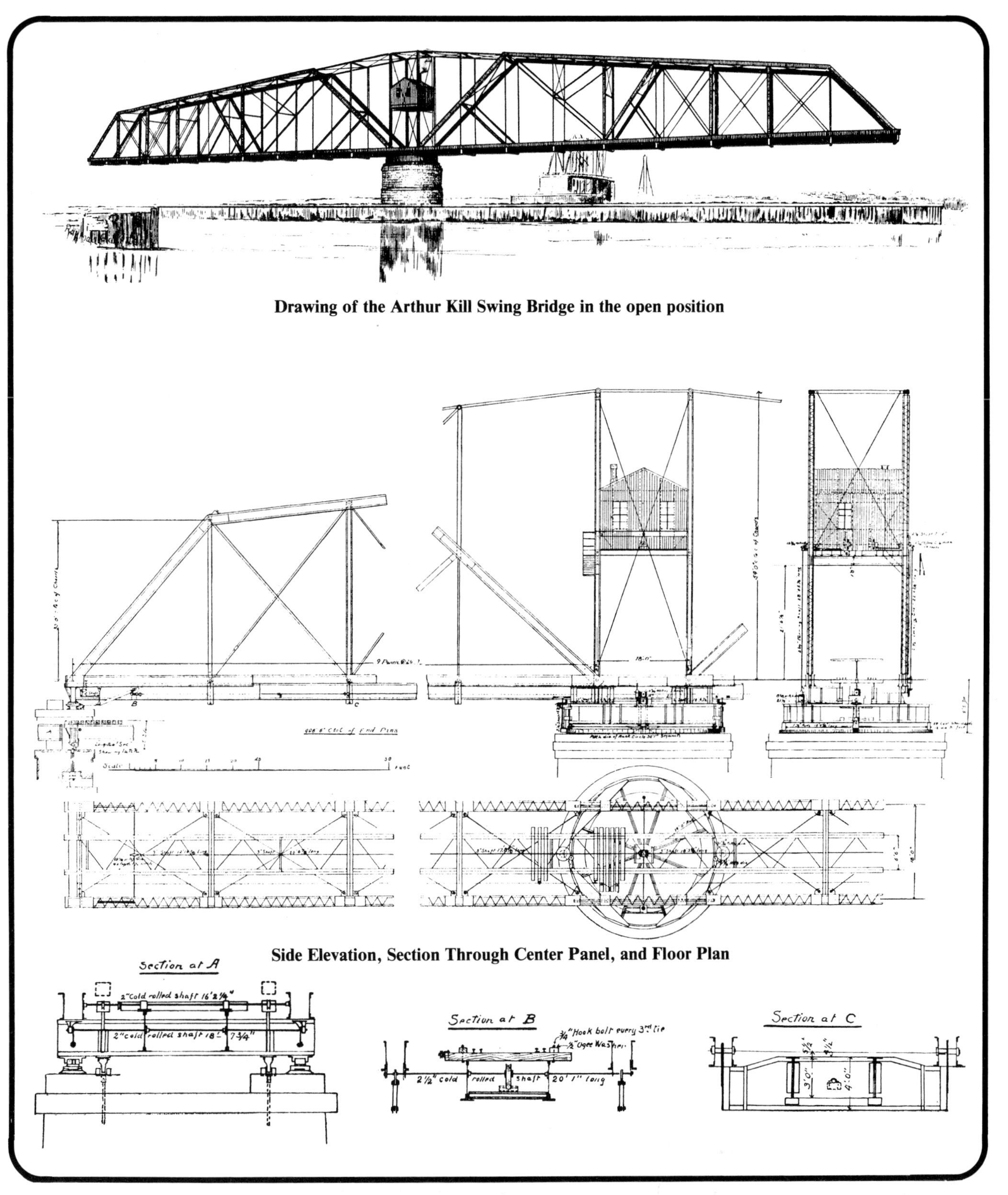

Drawing of the Arthur Kill Swing Bridge in the open position

Side Elevation, Section Through Center Panel, and Floor Plan

In 1888 the Staten Island Rapid Transit built a 496 foot swing bridge along Staten Island Sound (Arthur Kill) at the south end of Newark Bay, New Jersey, connecting Staten Island by rail for the first time. Fixed spans were located at both ends of the swing section of the bridge. The swing section of the structure was rim-bearing, the weight being distributed on eight equidistant points on a drum turntable. The bridge was operated from a house high above the tracks in the center of the swing portion of the bridge. This structure was replaced in 1959 by a 550 foot long lift bridge (see pages 120 and 123).

CONCRETE ARCH BRIDGE

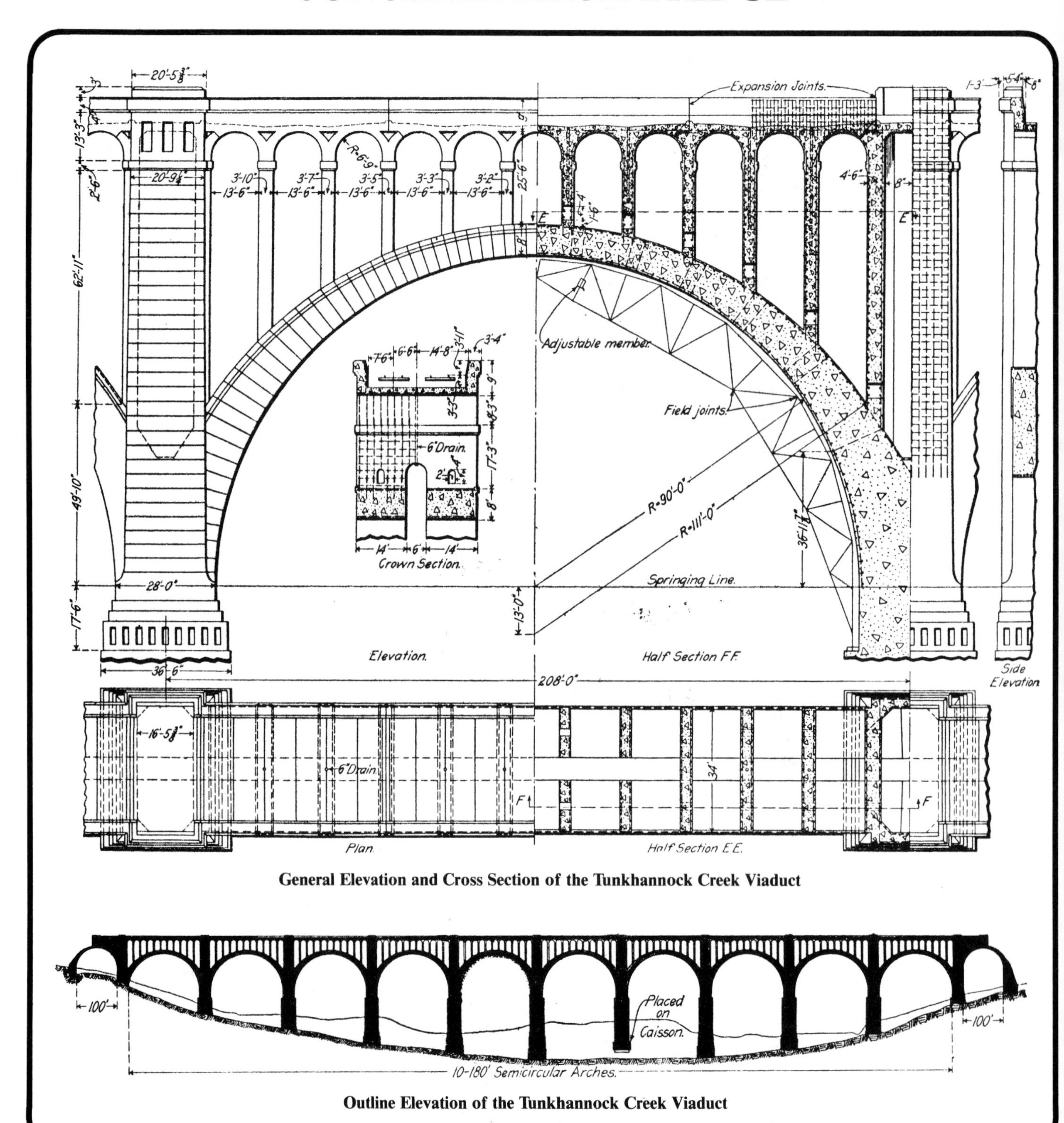

General Elevation and Cross Section of the Tunkhannock Creek Viaduct

Outline Elevation of the Tunkhannock Creek Viaduct

The use of concrete in railroad bridge building was slow to evolve due to the strict engineering standards set by bridge engineers. The first reinforced concrete bridges were monuments to the bridge building art. The Delaware, Lackawanna & Western Railroad's Clark's Summit Cut-Off included the Tunkhannock Creek Viaduct (see pages 169-171). This plan shows the general elevation and a cross section of this magnificent concrete arch structure as reproduced in the February 5, 1915 issue of the *Railway Age Gazette*.